The Kalman Filter
Introduction

Kalman Filter

Copyright
Copyright © 2021 by A. Maclean
All rights reserved.

Table of Contents

1 Introduction 11
   1.1 Overview 11
   1.2 Prerequisites 13
   1.3 Notation 13
2 Study Structure 16
3 Linear Systems 17
   3.1 Solving a Linear Differential Equation 19
   3.2 Modelling a Linear System 20
   3.3 Stability of a Linear System 21
   3.4 Lyapunov Stability 24
      3.4.1 Example: Lyapunov Criterion 27
4 Probability 29
   4.1 Random Variable 29
   4.2 Mean and Variance 29
   4.3 Probability Distribution Function 30
   4.4 Probability Mass Function 30
   4.5 Probability Density Function 30
   4.6 Cumulative Distribution Function 32
   4.7 Covariance 33
      4.7.1 Standardised Random Variables 34
      4.7.2 Covariance vs Correlation 34
      4.7.3 Autocorrelation 35
   4.8 Uniform Distribution 35
   4.9 Gaussian Distribution 37
      4.9.1 Body and Tail of a Distribution 40
   4.10 Multivariate Gaussian Distribution 41

- 4.11 Random Vector 41
- 4.12 Random Process 42
  - 4.12.1 Strict Sense Stationary 43
  - 4.12.2 Wide Sense Stationary 43
  - 4.12.3 White Noise 43
  - 4.12.4 The Statistical Ensemble 44
  - 4.12.5 Ergodicity 45
  - 4.12.6 Example: Mean Ergodicity 46
- 4.13 Bayes Theorem 46
  - 4.13.1 Example: Balls Drawn from a Bag 47
- 4.14 The Law of Total Probability 48
  - 4.14.1 Example: Law of Total Probability 49
- 4.15 Likelihood 49
  - 4.15.1 Maximum Likelihood Estimator 50
  - 4.15.2 Example: Gaussian 51

5 Least Squares Methods 53

6 Digital Filter 55
- 6.1 Non-Recursive Filter 56
- 6.2 Recursive Filter 57
- 6.3 Stochastic Processes 58

7 Markov Processes 59
- 7.1 Markov Chain 59
- 7.2 Stochastic Matrix 61
- 7.3 Hidden Markov Model 62
  - 7.3.1 The Kalman Filter Viewed as a HMM 65
- 7.4 Chapman-Kolmogorov Equations 65
  - 7.4.1 Example: C-K Equations 66

Kalman Filter

8	Alpha-Beta Filter	69
9	Estimation	71
9.1	Definition of Terms	71
9.1.1	Estimation Algorithm	71
9.1.2	Estimate	71
9.1.3	Accuracy	71
9.1.4	Bias	71
9.1.5	Precision	72
9.1.6	Repeatability and Reproducibility	72
9.2	Recursive Estimation	73
9.3	Linear Mean-Squared Estimator	74
9.4	Fisher Information	75
9.5	Cramer Rao Lower Bound	77
10	Linear Dynamic Systems	80
10.1	State	80
10.2	State Space	80
10.3	Matrix Rank	80
10.4	Observable Systems	81
10.5	Controllable Systems	83
11	The Kalman Filter	84
11.1	Random Noise	84
11.2	Kalman-Bucy Filter	85
11.3	Discrete Time Kalman Filter	88
11.3.1	Derivation of Scalar KF	89
11.3.1.1	Assumptions	89
11.3.1.2	Calculation of Alpha	90
11.3.1.3	Calculation of Beta	92

- 11.3.1.4 Solving Quadratic Equation for Beta............ 94
- 11.3.1.5 Solving for Alpha.. 95
- 11.3.1.6 The Scalar Kalman Filter............................... 95
- 11.3.2 Derivation of Vector KF Equations....................... 96
- 11.4 Kalman Predictor.. 100
  - 11.4.1 The Scalar Kalman Predictor Equations 100
  - 11.4.2 The Vector Kalman Predictor............................. 101
- 11.5 Estimation and Prediction 101
- 12 Implementation 103
  - 12.1 Digital Systems... 103
  - 12.2 Roundoff Error.. 103
  - 12.3 Positive Definite Matrix 104
  - 12.4 Positive Semi-definite Matrix..................................... 104
  - 12.5 Matrix Inversion 104
  - 12.6 Errors, Innovations, and Residuals 105
  - 12.7 Cholesky Decomposition.. 106
    - 12.7.1 Example: Cholesky Decomposition..................... 106
  - 12.8 Covariance Matrix Symmetry..................................... 108
- 13 Examples... 110
  - 13.1 A Note on Software 110
  - 13.2 Random Number Generation...................................... 111
    - 13.2.1 Gaussian Pseudorandom Numbers....................... 112
    - 13.2.2 The Box-Muller Method...................................... 112
  - 13.3 Position Estimate 118
    - 13.3.1 Description... 118
    - 13.3.2 Results.. 118
    - 13.3.3 Code 119

13.4 Biased Measurement Failure 123
13.4.1 Description 124
13.5 Biased Measurement 124
13.5.1 Description 125
13.5.2 Results 126
13.5.3 Code 127
13.6 Falling Body – with Driving Force 136
13.6.1 Description 136
13.6.2 Results 137
13.6.3 Code 138
13.7 Radar System 147
13.7.1 Description 147
13.7.2 Results 148
13.7.3 Code 149
13.8 Utility Functions 160
13.8.1 Gaussian Random Noise 161
13.8.2 Matrix Inversion 162
14 Further Reading 164
14.1 Extended Kalman Filter 164
14.2 Square Root Kalman Filter 164
14.3 Noise Models 166
14.4 Adaptive Kalman Filter 166
14.5 Data Fusion 166
14.6 Numerical Methods 166
15 Appendices 167
15.1 Matrix Identities 167
15.1.1 The Jacobian 167

- 15.1.2 Inversion Relation 168
- 15.2 Matrix Differentiation 168
  - 15.2.1 Simple Product 169
  - 15.2.2 Bilinear Form 169
  - 15.2.3 Differentiate Matrix wrt Scalar 169
  - 15.2.4 Differentiate Inverse 170
- 15.3 Curvature 170
  - 15.3.1 Motivation 170
  - 15.3.2 Curvature of a Planar Curve 173
  - 15.3.3 Example: Curvature of a Parabola 175
- 15.4 Integral for a Mean Ergodic Process 176
- 15.5 Integrals of Odd and Even Functions 178
- 15.6 Gaussian Integrals 180
- 16 References 183
- 17 Index 184

Abbreviations

AR	Autoregressive
ARMA	Autoregressive Moving Average
AWGN	Additive White Gaussian Noise
CDF	Cumulative Distribution Function
DSP	Digital Signal Processing
EKF	Extended Kalman Filter
FIR	Finite Impulse Response
HMM	Hidden Markov Model
iff	if and only if
IID	Independent and Identically Distributed
IIR	Infinite Impulse Response
KF	Kalman Filter
LHS	Left Hand Side
LQE	Linear Quadratic Estimation
LS	Least Squares
LTI	Linear Time Invariant
LU	Lower Upper
MA	Moving Average
ML	Maximum Likelihood
N(a, b)	Normal pdf with mean a and variance b
pdf	Probability Density Function
PDF	Probability Distribution Function
pmf	Probability Mass Function
PRNG	Pseudo-Random Number Generator

Kalman Filter

RHS	Right Hand Side
RMS	Root Mean Square
SISO	Single Input Single Output
SSS	Strict Sense Stationary
STM	State Transition Matrix
U(a, b)	Uniform pdf with constant value in range [a, b]
UD	Upper Triangular Diagonal
vs	Versus
wrt	With respect to
WSS	Wide Sense Stationary

1 Introduction

This section provides an introductory overview of the Kalman filter, its uses, and applications. This is followed by a list of assumed prerequisites and then a discussion of the various notations in use. Finally, a summary is provided of the notation used in this book.

1.1 Overview

The Kalman filter is an optimum recursive linear mean-squared estimator. Its function is the determination of the best estimate of a signal in the presence of noise. Specifically, the Kalman filter has the following properties:

> Optimum – The Kalman filter is optimal in the sense that it provides the statistical estimate with the minimum error covariance based on all the available observations at that time. The system and measurement noise are assumed to be Gaussian white (uncorrelated) noise.
>
> Recursive – The Kalman filter is a recursive estimator. The current state is computed from the previous estimate and the current measurement.
>
> Linear – The Kalman filter is based on linear dynamical systems. It is an estimator of the internal state of the linear system.
>
> Mean-square -The Kalman filter is the best possible linear estimator in the sense that it minimises the mean square error.
>
> Estimator - An estimation algorithm, such as a Kalman filter, uses a series of noisy measurements, taken over past times, to provide an improved estimate of unknown variables.

A Kalman filter estimates a system state vector in the presence of system and measurement noise. The following terms are fundamental to such an estimator.

Kalman Filter

Noise – The Kalman filter estimates the state of the system in the presence of statistical system noise and measurement noise.

State – The state of a system is a collection of variables that describe the behaviour of the system. Examples are position and velocity.

Dynamics – A dynamical system is a system in which a function describes the time dependence of the system. An example is a pendulum.

Stochastic – A stochastic (random) process is defined by a random probability distribution. In the case of the Kalman filter, the randomness is introduced by system noise and measurement noise.

A generic Kalman filter and observed system are illustrated in Figure 1. The system (sometimes known as the plant) under observation is subject to external noise. The measurement of some system parameters is subject to measurement noise. The Kalman filter uses a model of the system dynamics together with the noisy measurement to provide an estimate of the internal state of the system.

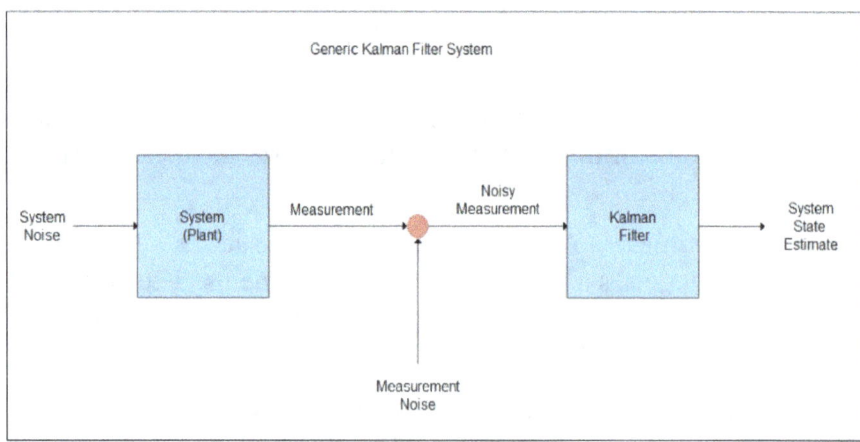

Figure 1 – Generic System and Kalman Filter

Kalman Filter

The terms listed above are examined in detail in the following chapters.

1.2 Prerequisites

The following prerequisites are assumed:

> Linear Algebra – matrices and vectors.
>
> Probability – Gaussian distribution, Bayes' theorem.
>
> Laplace transform.
>
> Vector calculus.
>
> Linear Differential Equations.

These are all discussed in [1].

1.3 Notation

There are a number of common notations used in the literature. For future reference, Table 1-1 summarises the notation used herein and lists some of the common alternatives. The three columns detail:

1. Our notation (the variable)
2. The name of each quantity
3. Common alternative notations

Kalman Filter

Variable	Name	Common Alternatives
x	State vector	
y	Measurement vector	
Φ(k)	System transition matrix	A, F
H(k)	Measurement matrix	C
Q(k)	System noise covariance matrix	
R(k)	Measurement noise covariance matrix	
P(k)	Error covariance matrix	
K(k)	Kalman gain matrix	
$\hat{x}(k\|k)$	Best estimate of **x**(k) given data at time k (filtered estimate)	$\hat{x}(k)$
$\hat{x}(k\|k-1)$	Best estimate of **x**(k) given data at time k − 1 (predicted estimate)	$\hat{x}(k-1)$, $\hat{x}^-(k)$
$\hat{P}(k\|k)$	Covariance matrix of filtered estimate	$\hat{P}(k)$
$\hat{P}(k\|k-1)$	Covariance matrix of predicted estimate	$P_1(k), P^-(k)$
Γ(k)	Input matrix	B, G
u(k)	Forcing function	**y**(k)
w(k)	Process noise	
v(k)	Measurement noise	
k	A time instant	

Table 1-1 – Notation

Kalman Filter

The Kalman filter equations are summarised here:

Extrapolation:

$$\hat{x}(k|k-1) = \Phi(k-1)\,\hat{x}(k-1|k-1)$$

$$P(k|k-1) =$$
$$\Phi(k-1)P(k-1|k-1)\Phi(k-1)^T + Q(k-1)$$

Measurement:

Measure **y**(k)

Update:

$$K(k) =$$
$$P(k|k-1)\,H(k)^T\{H(k)P(k|k-1)H(k)^T + R(k)\}^{-1}$$
$$P(k|k) = [I - K(k)H(k)]\,P(k|k-1)$$
$$\hat{x}(k|k) = \hat{x}(k|k-1) + K(k)\,[y(k) - H(k)\hat{x}(k|k-1)]$$

2 Study Structure

The following sections review the foundation material required for the rest of the book. This includes:

- Linear Systems
- Probability
- Least Squares Methods
- Digital Filters
- Markov Processes
- The α-β Filter
- Estimation

The Kalman filter sections follow and they include:

- Linear Dynamic Systems
- The Kalman Filter
- Implementation
- Examples
- Further Reading

The Appendices introduce required material which does not fit naturally into the main text.

A good overview of related material is given in [2].

3 Linear Systems

The reasons for studying linear systems are that:

1. Practical engineering systems are often linear, at least to a good approximation.
2. Exact solutions to linear systems of equations are easily found using standard methods.
3. For a linear system, the superposition principle holds.
4. Analytical solutions to non-linear systems are usually unavailable.
5. Linear systems can be used to approximate non-linear systems.

The superposition principle is defined as follows. If a system with inputs x(t) and outputs y(t) is described by the relation

$$y(t) = Hx(t)$$

then the following conditions apply for inputs $x_1(t)$ and $x_2(t)$

$$y_1(t) = Hx_1(t)$$
$$y_2(t) = Hx_2(t)$$
$$\alpha\, y_1(t) + \beta\, y_2(t) = H\{\alpha x_1(t) + \beta x_2(t)\}$$

for any scalars α, β.

Another way to view the superposition principle, in more intuitive terms, is that linear systems are important because for a linear system with forcing function $f_1(t)$ and response function $r_1(t)$ and then with a forcing function $f_2(t)$ and response function $r_2(t)$ the forcing function $f_1(t) + f_2(t)$ produces a response function $r_1(t) + r_2(t)$. This can be summarised as:

For a linear system such that

$$f_1(t) \rightarrow r_1(t)$$
$$f_2(t) \rightarrow r_2(t)$$

then

$$f_1(t) + f_2(t) \rightarrow r_1(t) + r_2(t)$$

Kalman Filter

The superposition of individual forcing functions results in a response that is the superposition of the individual responses. The principle of superposition is a necessary condition for a system to be linear.

This means that

1. No forcing function affects the response of any other forcing function.
2. There is no interaction between the responses caused by different forcing functions.
3. The combined effect of a number of forcing functions on a linear system can be found by the separate determination of the individual response functions and then forming a combination of the individual responses to determine the overall system response function.

In addition, n identical response functions have the effect that

$$nf(t) \rightarrow nr(t)$$

This is referred to as the principle of homogeneity.

A system is linear only if both the principle of superposition and the principle of homogeneity are obeyed.

Linear systems are modelled by linear differential equations. It is often necessary to covert a continuous linear system into a discrete linear system for implementation on a digital computer. This is achieved by solving the differential equation describing the linear system, and then finding a suitable discrete approximation to the system.

A linear system is stationary if, in addition to the above considerations, when a periodic forcing function with frequency f is applied to the system, then the steady state response of the system is also periodic with frequency f. The linear system is stationary if

$$f(t - T) \rightarrow r(t - T)$$

for an arbitrary time-delay T. The properties of the system do not change with time.

Kalman Filter

3.1 Solving a Linear Differential Equation

We consider a first order process and solve the equation in the standard way. The general first order equation describing the system is

$$\frac{dy}{dx} + a(x)y = f(x)$$

To convert the lhs into a product that can be differentiated we introduce an integrating factor u(x) defined, using the Leibnitz rule, as

$$\frac{d(uy)}{dx} = u\frac{dy}{dx} + \frac{du}{dx}y$$

Then multiply by u(x) to give

$$u(x)\frac{dy}{dx} + u(x)a(x)y = u(x)f(x)$$

For the lhs to be an exact derivative we require that

$$\frac{du}{dx} = u(x)a(x)$$

Hence

$$\int \frac{du}{u} = \int a(x)dx$$

giving

$$\ln(u) = \int a(x)dx$$
$$u(x) = \exp\left(\int a(x)dx\right)$$

From the Leibnitz rule above

$$\frac{d}{dx}\left(y\exp\int a(x)dx\right) = \frac{dy}{dx}\exp\int a(x)dx + ya(x)\exp\int a(x)dx$$

$$\frac{d}{dx}\left(y\exp\int a(x)dx\right) = \left[\frac{dy}{dx} + a(x)y\right]\exp\int a(x)dx$$

$$\frac{d}{dx}\left(y\exp\int a(x)dx\right) = f(x)\exp\int a(x)dx$$

Kalman Filter

Hence

$$y(x) = e^{-\int a(x)dx}[\int f(x) e^{\int a(x)dx} dx + C]$$

In the next section we use this solution to produce a discrete model of a continuous system.

3.2 Modelling a Linear System

Consider a system given by the following equation:

$$\dot{x}(t) = \Phi x(t) + \Gamma u(t)$$

where Φ, Γ are constants. The system is:

- linear
- a first order differential equation
- single input, single output (SISO)
- time invariant (constant Φ, Γ)
- deterministic (noise free)

The function u(t) is an input forcing function. The initial condition is given by $x(0) = x_0$. The solution to the equation is found as in the previous section and is given by

$$x(t) = e^{\Phi t} x_0 + \int_0^t e^{\Phi(t-\tau)} \Gamma u(\tau) d\tau$$

A discrete approximation of this system is necessary to implement a model of the system on a digital computer. Hence, we must determine the values of x(t) at discrete intervals given by T seconds. At time T we obtain

$$x(T) = e^{\Phi t} x_0 + \int_0^T e^{\Phi(T-\tau)} \Gamma u(\tau) d\tau$$

and a change of variables yields

$$x(T) = e^{\Phi t} x_0 + \int_0^T e^{\Phi \tau} \Gamma u(T - \tau) d\tau$$

Similarly

$$x(2T) = e^{\Phi t} x(T) + \int_0^T e^{\Phi \tau} \Gamma u(2T - \tau) d\tau$$

Kalman Filter

and the sequence continues to give

$$x(kT) = e^{\Phi t}x((k-1)T) + \int_0^T e^{\Phi \tau}\Gamma u(kT-\tau)d\tau$$

Choose T to be small enough such that u(t) does not change significantly during that time interval, and then, to good approximation,

$$x(kT) = e^{\Phi t}x((k-1)T) + \Gamma u(kT)\int_0^T e^{\Phi \tau}d\tau$$

$$x(kT) = A(x((k-1)T) + Bu(kT)$$

where the constants A and B are seen to be given by

$$A = e^{\Phi t}$$

$$B = \Gamma\int_0^T e^{\Phi \tau}d\tau = \frac{\Gamma}{\Phi}[e^{\Phi t} - 1]$$

The constants A and B are computed offline.

We now have a difference equation with constant coefficients and this can be modelled digitally. The general form of this equation is (suppressing T)

$$x(k) = Ax(k-1) + Bu(k)$$

This is an important equation because this is the common form of the system equation in the treatment of the Kalman filter.

3.3 Stability of a Linear System

This section assumes a knowledge of the Laplace transform.

The stability of a linear stationary process can be determined by examination of the system's impulse response. The forcing function is a delta function (the impulse) and the response is examined to establish the stability of the system with respect to this simplest of forcing functions.

Consider the linear differential equation given by

$$\alpha \dot{x} + x = f(t) \qquad x(0) = 0$$

A Laplace transform is used as follows to convert the differential equation to an algebraic equation.

Kalman Filter

$$\mathcal{L}(\alpha \dot{x} + x) = \mathcal{L}f(t)$$
$$\alpha \mathcal{L}(\dot{x}) + \mathcal{L}(x) = \mathcal{L}f(t) \qquad \text{Linear}$$
$$\alpha[sx(s) - x(0)] + x(s) = f(s)$$

Take the Laplace transform

$$(\alpha s + 1)x(s) = f(s) + \alpha x(0)$$

$$x(s) = \frac{f(s)}{\alpha s + 1} + \frac{\alpha x(0)}{\alpha s + 1}$$

The solution is given by the inverse Laplace transform

$$x(t) = \mathcal{L}^{-1}\left(\frac{f(s)}{\alpha s + 1}\right) + x(0)(1 - e^{-\frac{t}{\alpha}})$$

The RHS is the usual sum of particular integral and complimentary solution, respectively.

The stability is determined by the nature of the forcing function. If the forcing function is a delta function, then we have

$$x(s) = \frac{\mathcal{L}\delta(0)}{\alpha s + 1}$$

The Laplace function of the delta function is one and so

$$\mathcal{L}\delta(0) = 1 \qquad \text{Standard result}$$

The delta function can be thought of as the derivative of the step function and we can motivate this result as follows

$$\mathcal{L}\delta(0) = \mathcal{L}\frac{dU(t)}{dt} = s\mathcal{L}U(t) = s\left(\frac{1}{s}\right) = 1$$

Hence

$$x(s) = \frac{1}{\alpha s + 1}$$

and then

$$x(t) = \mathcal{L}^{-1}x(s) = e^{-t/\alpha} = e^{Spole\, t}$$

The system stability is related to the position of the pole in the complex s-plane. For stability we require the exponent to be negative, and so

Kalman Filter

$$Re(s_{pole}) < 0$$

In this case the exponent is negative and the impulse response is bounded and decreasing in magnitude.

This approach to solving linear differential equations can be applied to higher order linear systems. For an nth order system to be stable, all poles of the system must have negative real part. In other words, they must lie in the left-hand half of the s-plane. This is summarised in Figure 2.

Kalman Filter

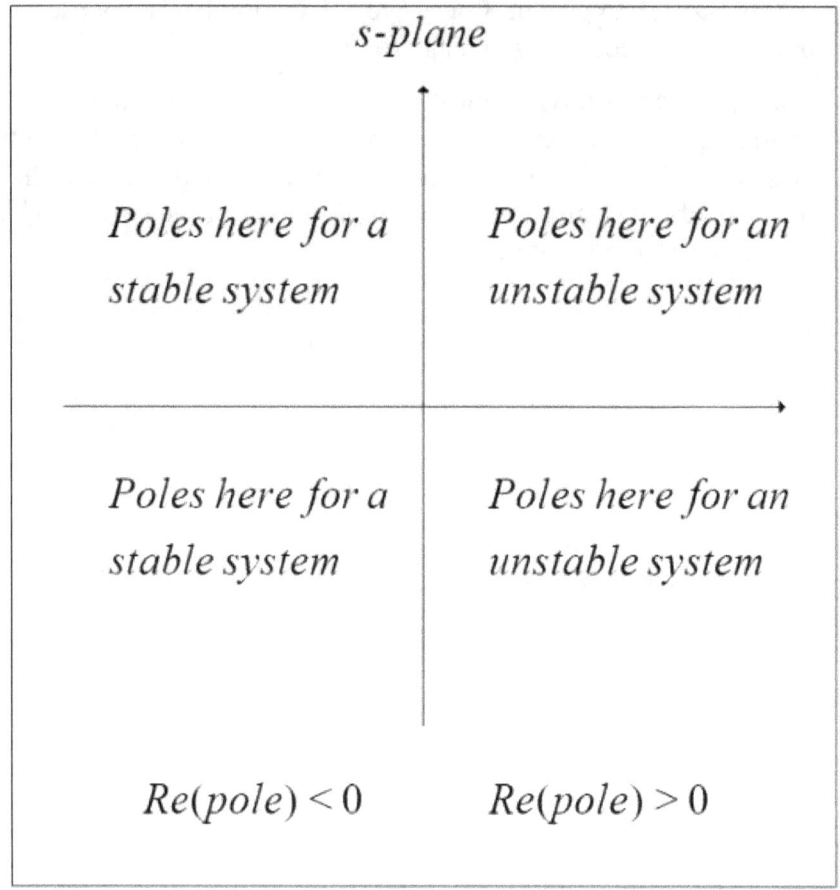

Figure 2 – Stable Poles of a Linear System

3.4 Lyapunov Stability

The theme of stability continues with the Lyapunov approach. We begin with a motivating example and then state the general theorem. The example is a familiar one. It is that of a ball rolling down a slope, for example, a ball in a bowl. The system is illustrated in Figure 3.

There are two cases.

Kalman Filter

In case A the ball rolls down from the left and it is observed that the direction of motion is in the positive direction. The gradient of the slope of the bowl is negative. The product of the signs of these two quantities is negative.

In case B the ball rolls down from the right and it is observed that the direction of motion is in the negative direction. The gradient of the slope of the bowl is positive. The product of the signs of these two quantities is negative.

In general

$$direction\ of\ motion\ \times gradient\ of\ motion < 0$$

In both cases the trend is to force the ball toward the equilibrium point, as illustrated.

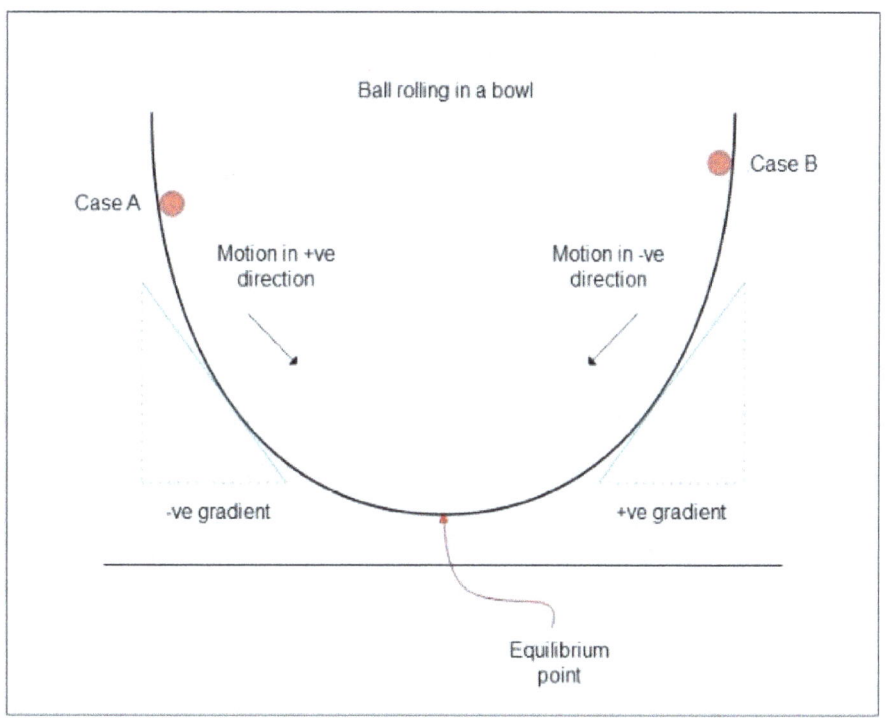

Figure 3 – Lyapunov Stability

Now, to generalise this idea, consider a system with a dissipative

Kalman Filter

force
$$m\ddot{x} = -\beta \dot{x} \qquad \beta > 0$$

The kinetic energy is given by
$$E(t) = \frac{1}{2}mv^2$$

and differentiating
$$\dot{E}(t) = mv\dot{v} = mv\left[\frac{-\beta\dot{x}}{m}\right] = -\beta v^2 < 0$$

Hence, $E(t) \geq 0$ and $\dot{E}(t) < 0$ and so $E(t)$ is greater than zero and is decreasing. Hence, we can deduce that
$$E(t) \to 0 \text{ as } t \to \infty$$
$$v(t) \to 0 \text{ as } t \to \infty$$

and this implies that the system is stable. The ball will eventually be at rest at the equilibrium point.

The problem now, for any system, is to identify a suitable Lyapunov function. Such a choice is often a quadratic form of the type
$$L(x) = x^T V x$$

where V is real and symmetric and $L > 0$.

The above motivation permits us to state without proof the Lyapunov stability criterion is as follows:

For a system $\dot{x} = f(x)$, a point x_s for which $f(x_s) = 0$ is called a singular point.

If the origin is a singular point, then it is stable if a Lyapunov function L(x) can be found with the following properties:

1. $L(x) > 0 \text{ for all } x \neq 0$
2. $\frac{dL}{dt} \leq 0$ for all x
3. If $\frac{dL}{dt}$ is zero only at the origin, the origin is asymptotically stable (all trajectories x(t) tend toward the singular point as $t \to \infty$).

Kalman Filter

3.4.1 Example: Lyapunov Criterion

The following example illustrates the Lyapunov criterion. Consider a system given by

$$\dot{x} = -f(x) \qquad f(0) = 0 \qquad xf(x) > 0$$

where f is a function as illustrated in Figure 4.

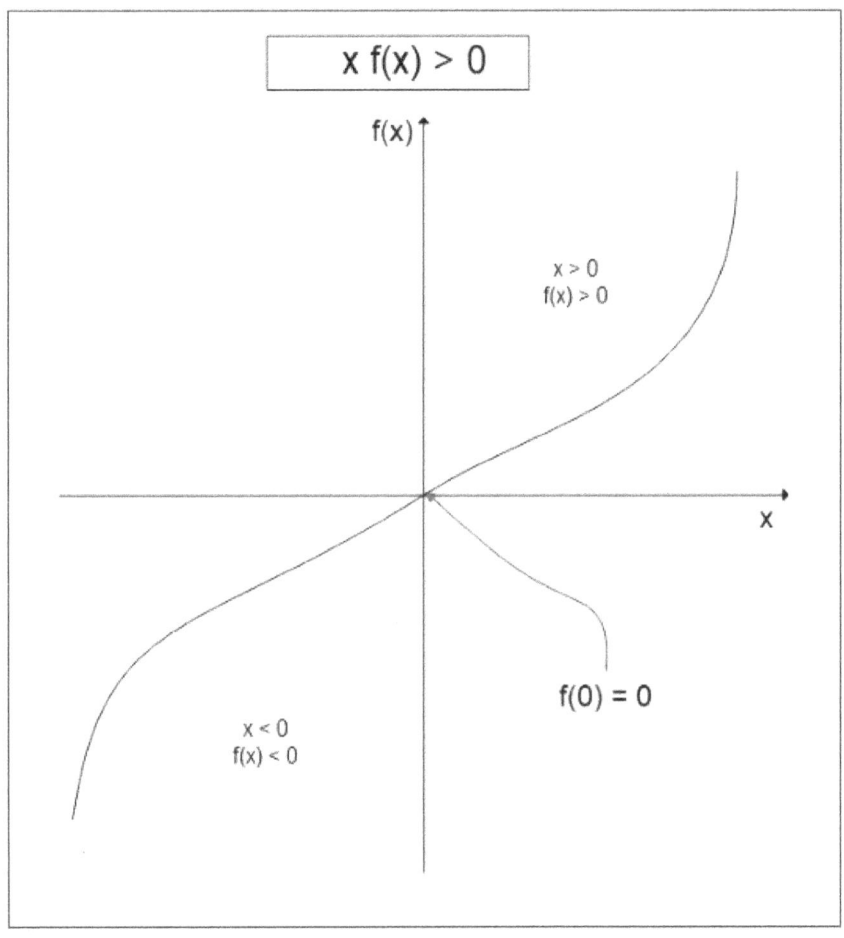

Figure 4 – Example Forcing Function

Kalman Filter

Note that $xf(x) > 0$ always.

Choose the Lyapunov function to be

$$L(x) = \frac{1}{2}x^2 > 0 \quad \forall\, x \neq 0$$

Then we have

$$\frac{dL}{dt} = x\dot{x} = -xf(x) \leq 0 \quad \forall\, x$$

and according to the Lyapunov criteria above, the system is asymptotically stable.

4 Probability

Some terms used in probability theory are defined below.

Marginal Probability is the probability of an event occurring, unconditionally. For example, the probability of a coin turning up heads is 0.5. It does not depend on any other event. If the probability of an event is 1 then that event will happen with certainty. Conversely, if the probability is zero then the event will never occur.

Conditional Probability is the probability that an event will occur, given that some other event first occurred.

The sum of the probabilities of all outcomes must be equal to one.

4.1 Random Variable

A random variable takes different numerical values according to the outcome of some operation. The results are unpredictable, such as the outcome of rolling a die. In this case the result is one member of the set $\{1, 2, 3, 4, 5, 6\}$. A random variable can be discrete or continuous. The value of the random variable can be represented by a probability distribution. For a discrete system this could be a coin toss $\{Heads, Tails\}$. For a continuous system it could be a value taken from a Gaussian distribution $p(x)$.

4.2 Mean and Variance

The values in a data set can be characterised in a number of ways. The mean, median, and mode are measures of the central values of the set. The range of the values is the difference between the highest and lowest values in the set.

The median value of the set is the value that is the middle number when the set is ordered. If the set has an even number of elements, then the median is the average of the two middle numbers.

The mode is the value that occurs most often in the data set.

Kalman Filter

The mean of the data in the set is average value. Namely the sum of all the numbers in the set divided by the number of values in the set.

The variance of the data is found by finding the mean, as above. Then find the squared difference from the mean for each value. Sum all the squared differences and divide by the sum by the number of values.

The variance does not have the same units as the mean and so another useful measure is the standard deviation. This is the square root of the variance.

The standard deviation and the variance provide a measure of the spread of the values in the set.

4.3 Probability Distribution Function

In general, a probability distribution is a list of outcomes and their associated probabilities.

A function that represents a discrete probability distribution is called a probability mass function. It is a list of each possible values taken by the quantity of interest together with the probability that a given value is obtained in a single trial.

A function that represents a continuous probability distribution is called a probability density function.

4.4 Probability Mass Function

The probability mass function (discrete probability function) is a function that gives the probability that a discrete random variable is equal to some value. It is the discrete counterpart of the continuous probability density function. A pmf is often represented by a histogram.

4.5 Probability Density Function

For discrete data, the number of items in a given range can be plotted as a histogram. A similar concept for continuous variables

Kalman Filter

is the probability density function (pdf). The pdf is a graph showing probability density of a value versus the (continuous) value. The pdf is non-negative. The total area under the curve must be equal to one, that is, the probability of attaining some value in the range must be one.

The probability of attaining a value between x_1 and x_2 is given by

$$P(1,2) = \int_{x_1}^{x_2} p(x)dx$$

This is illustrated in Figure 5 for the case of the exponential distribution.

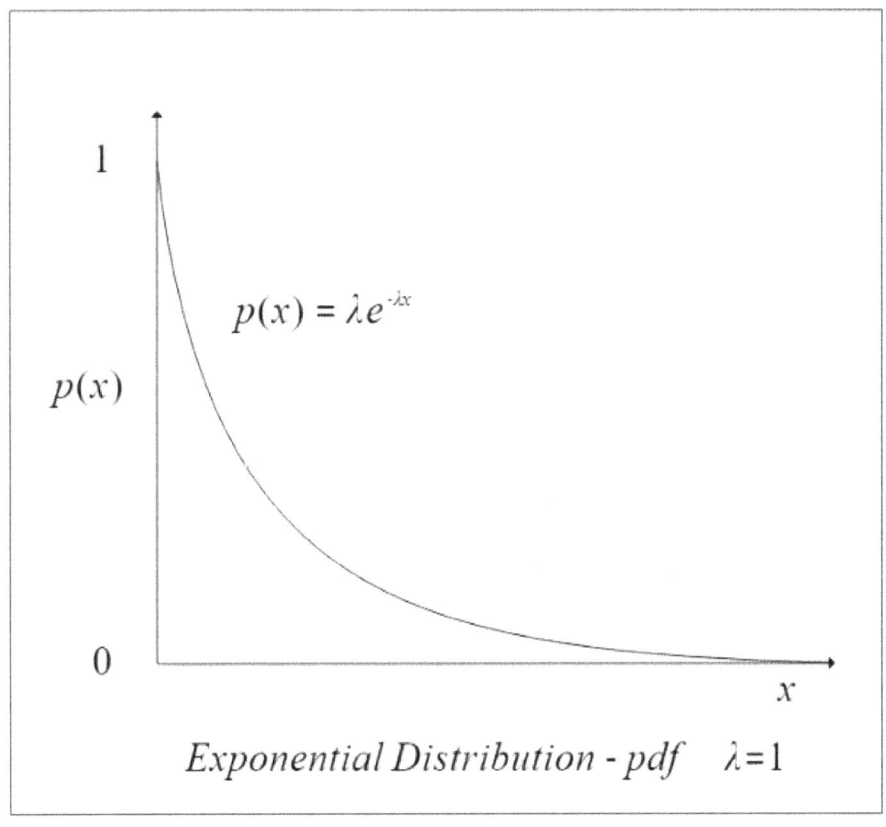

Figure 5 – Probability Density Function

Kalman Filter

4.6 Cumulative Distribution Function

The Cumulative Distribution Function (CDF) of a random variable X is the probability that an X value will be less than or equal to some specified value, x. That is

$$F(x) = \Pr(X \leq x)$$

In terms of the pdf (p) this is given by

$$F(x) = \int_{-\infty}^{x} p(t)dt$$

F(x) is a monotonically increasing function and $F(x) \to 1$ as $x \to \infty$ and $F(x) \to 0$ as $x \to -\infty$. See *Figure 6* for the case of the exponential distribution.

The CDF is calculated as follows:

$$P(x) = P[X \leq x] = \int_{-\infty}^{x} p(t)dt$$
$$= \int_{0}^{x} \lambda x^{-\lambda t} dt = \left[-e^{-\lambda t}\right]_{0}^{x}$$
$$= -e^{-\lambda x} - (-e^{0})$$
$$= 1 - e^{-\lambda x}$$

Kalman Filter

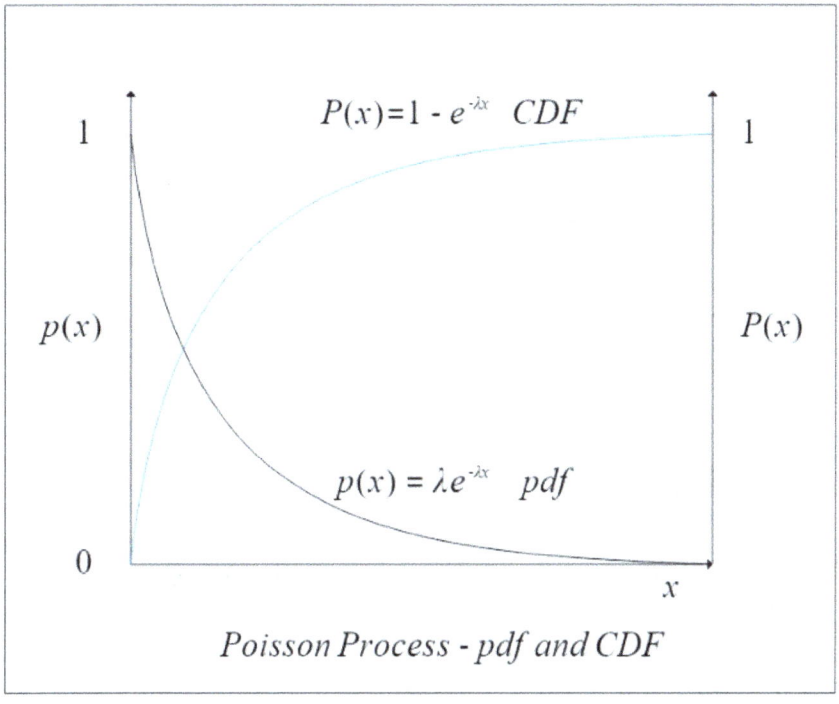

Figure 6 – Cumulative Distribution Function

4.7 Covariance

The ideas of mean and variance for a single process were introduced above. For the Kalman filter, this needs to be generalised to multivariable systems. If X and Y are jointly distributed random variable then the Covariance of X and Y is defined to be

$$Cov(X,Y) = E[(X - E(X))(Y - E(Y))]$$

and, in particular, the autocovariance is equal to the variance

$$Cov(X,X) = \sigma^2$$

The covariance measures the degree to which two random variables are related.

If **x** is a d-dimensional random vector then its covariance is

Kalman Filter

defined to be

$$Cov(x) = E((x - E(x))(x - E(x))^T)$$

$$Cov(x) = \begin{pmatrix} \sigma_1^2 & \cdots & \sigma_{1d} \\ \vdots & \ddots & \vdots \\ \sigma_{d1} & \cdots & \sigma_d^2 \end{pmatrix}$$

where:

$$\sigma_{xy} = E((x - E(x))(y - E(y)))$$

and the diagonal elements are the variance of each element of the vector.

4.7.1 Standardised Random Variables

If X is a random variable with mean and standard deviation μ and σ then both quantities are unit dependent. The random variable defined to be

$$X^* = (X - \mu)/\sigma$$

is called the standardised X. This has the following properties

$$E(X^*) = E\left(\frac{X}{\sigma} - \frac{\mu}{\sigma}\right) = \frac{1}{\sigma}(E(x) - \mu) = \frac{\mu}{\sigma} - \frac{\mu}{\sigma} = 0$$

$$Var(X^*) = Var\left(\frac{X}{\sigma} - \frac{\mu}{\sigma}\right) = \frac{1}{\sigma^2}Var(X) = \frac{\sigma^2}{\sigma^2} = 1$$

For X, the origin is changed to μ and the variable is scaled to units of the standard deviation. This is independent of the units of measurement.

4.7.2 Covariance vs Correlation

There can be confusion between the covariance matrix and the correlation matrix. Covariance values can be anything from 0 to infinity. If it is desired to normalise the values then the correlation matrix is used. This is defined to be

$$corr(Y, Y) = \frac{cov(X,Y)}{\sqrt{Var(X)Var(Y)}}$$

Kalman Filter

The correlation matrix has the limited range of values $-1 \leq corr(X,Y) \leq +1$. Each diagonal entry is +1 and the other entries are in the range from -1 to +1.

Note that autocorrelation is the cross-correlation of a signal with itself and autocovariance is the cross-covariance of a signal with itself.

The Kalman filter is concerned with the covariance matrix rather than the correlation matrix.

4.7.3 Autocorrelation

If a signal is correlated with a delayed copy of itself, the resulting function is called the autocorrelation. For a discrete signal, it is obtained by taking the expectation value of any two signal samples separated by k signal samples.

$$R_y(k) = E[y_n y_{n+k}]$$

The time separation k is called the lag.

For a continuous white noise process the signals will only be non-zero when the lag is zero. Hence the autocorrelation for a white noise signal has a strong peak at the origin (k=0) and is zero elsewhere. It is a delta function given by $\delta(0)$.

For a signal s(t) the continuous autocorrelation is

$$R_{ss}(\tau) - \int_{-\infty}^{+\infty} f(t+\tau) f(t) dt$$

4.8 Uniform Distribution

One of the simplest probability distributions is the Uniform probability distribution. The pdf of the uniform distribution is

$$u(x) = \frac{1}{B-A} \quad for \ A \leq x \leq B$$

The case where $A = 0, B = 1$ is called the standard uniform distribution. It is given by

$$U(x) = 1 \quad for \ 0 \leq x \leq 1$$

Kalman Filter

The uniform probability density function is illustrated in Figure 7. Note that in both cases the area under the curve is 1.

The probability of any event between the limits is given by the constant value $\frac{1}{B-A}$.

The uniform probability distribution is often represented as U(A, B).

The mean is given by

$$E(x) = \int_A^B xU(x)dx = \int_A^B \frac{x}{B-A} dx = \frac{B-A}{2}$$

The variance is given by

$$Var(x) = E(x^2) - E(x)^2 = \int_A^B \frac{x^2}{B-A} dx - \left(\frac{B-A}{2}\right)^2$$

$$= \frac{(B-A)^2}{12}$$

For the standard uniform distribution, the mean is $\frac{1}{2}$ and the variance is $\frac{1}{12}$.

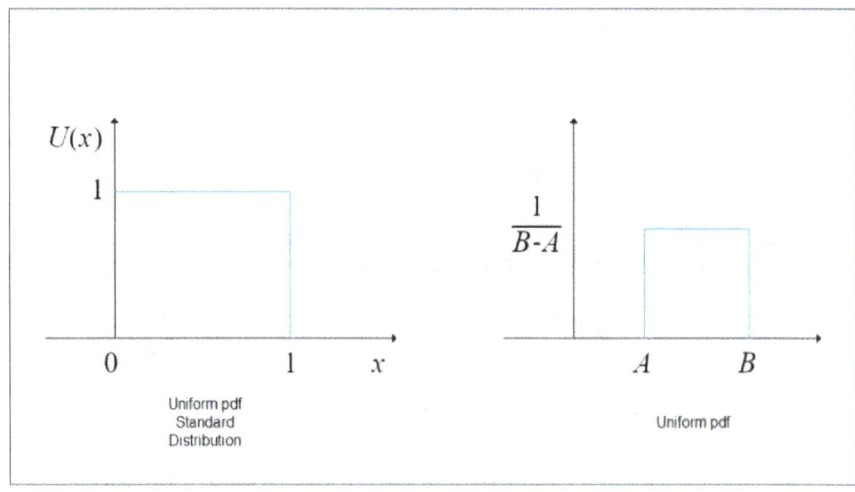

Figure 7 – Uniform Distribution

Kalman Filter

4.9 Gaussian Distribution

Consider the average value of the exponential function defined as follows:

$$\langle expik(y_N - \bar{x}) \rangle = \langle expik(x_1 + x_2 + \cdots + x_N - N\bar{x}/N) \rangle$$

The exponent is the difference between the mean of N terms less the value of the actual mean. The x_i are identically distributed, independent variables. The distribution of the variables is unspecified. Expanding the exponential, we obtain:

$$\langle expik(y_N - \bar{x}) \rangle =$$

$$\langle expik((x_1 - \bar{x}) + (x_2 - \bar{x}) + \cdots + (x_N - \bar{x}))/N \rangle$$

$$= \langle expik\left(\frac{(x_1-\bar{x})}{N}\right) expik\left(\frac{(x_2-\bar{x})}{N}\right) \cdots \frac{expik((x_N-\bar{x}))}{N} \rangle$$

$$= \langle expik\left(\frac{(x-\bar{x})}{N}\right) \rangle^N$$

This is because the mean of each of the N x_i is identical and it is denoted by $\bar{x}$. Expanding the exponential in a Taylor series gives

$$= \langle 1 + \frac{ik}{N}(x - \bar{x}) - \frac{k^2}{2N^2}(x - \bar{x})^2 + O\left(\left(\frac{k}{N}\right)^3\right) \rangle^N$$

The value of the mean is $\bar{x}$ and so we have

$$= \left[1 + 0 - \frac{k^2 \sigma_x^2}{2N^2} + O\left(\left(\frac{k}{N}\right)^3\right)\right]^N$$

Using the relation

$$\lim_{n \to \infty} \left(1 + \frac{x}{n}\right)^n = e^x$$

delivers the result (higher order terms vanish as $N \to \infty$).

$$\langle expik(y_N - \bar{x}) \rangle = e^{-k^2 \sigma_x^2 / 2N}$$

In terms of the probability density function we can express this as ($\bar{x} = constant$)

$$\int_{-\infty}^{+\infty} expik(y_N - \bar{x}) p(y_N - \bar{x}) dy_N = e^{-k^2 \sigma_x^2 / 2N}$$

Multiply both sides by $exp - ik(\alpha)$ and integrate over k

Kalman Filter

$$\int_{-\infty}^{+\infty} dk e^{-ik\alpha} \int_{-\infty}^{+\infty} expik(y_N - \bar{x}) p(y_N - \bar{x}) dy_N =$$

$$\int_{-\infty}^{+\infty} dk e^{-ik\alpha} e^{-k^2 \sigma_x^2 / 2N}$$

Integrating the lhs over k, and noting the delta function $\delta((y_N - \bar{x}) - \alpha)$, gives

$$p(\alpha) = \frac{1}{2\pi} \int_{-\infty}^{+\infty} dk e^{-ik\alpha} e^{-k^2 \sigma_x^2 / 2N}$$

Using the standard integral

$$\int_{-\infty}^{+\infty} dt e^{-i\alpha t} e^{-\beta t^2} = \sqrt{\frac{\pi}{\beta}} e^{-\frac{\alpha^2}{4\beta}}$$

we obtain

$$p(\alpha) = \sqrt{\frac{N}{2\pi\sigma^2}} e^{-\frac{\alpha^2 N}{2\sigma^2}}$$

Letting $\alpha = (x - \bar{x})$ and noting that the standard deviation of the sum is $\frac{\sigma}{\sqrt{N}}$ we obtain the Gaussian distribution

$$N(x - \mu) = \frac{1}{\sqrt{2\pi\sigma^2}} exp\left[-\frac{1}{2}(x - \mu)^2 / \sigma^2\right]$$

for the variable x with mean μ and standard deviation σ. The Gaussian distribution is illustrated in Figure 8.

In the limit as $N \to \infty$, the sum of N variables approaches the Gaussian distribution independent of the probability distribution of the variables. This is called the Central Limit Theorem. This theorem is central to the importance of the Gaussian distribution in the study of random processes. It means that many random processes are Gaussian.

Kalman Filter

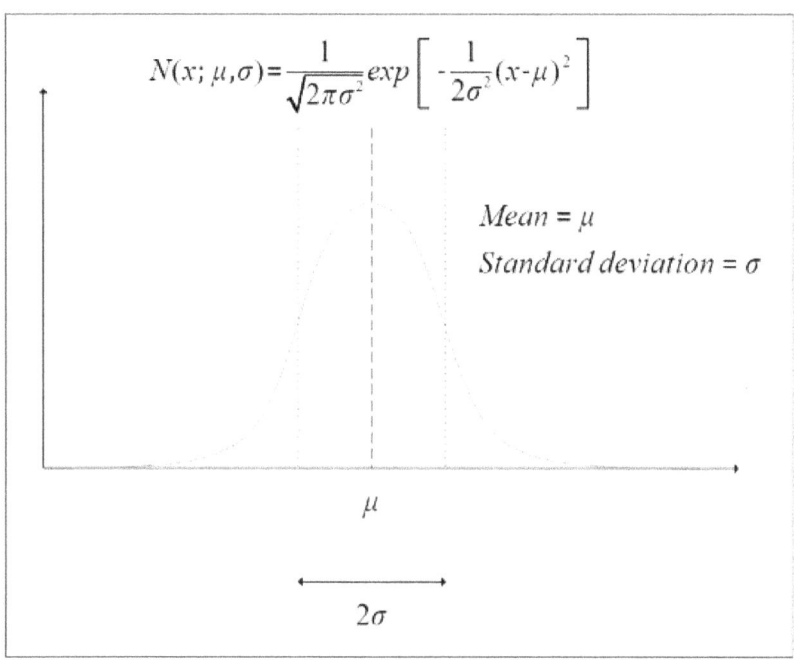

Figure 8 – Gaussian Distribution

In Figure 9 the probability of the value x lying between the values x_1 and x_2 is given by

$$P_{12} = \int_{x_1}^{x_2} N(x)dx$$

and the probability of the value x lying above the value x_1 is given by

$$P_{1\infty} = \int_{x_1}^{\infty} N(x)dx$$

and the mean value is

$$\mu = \int_{-\infty}^{+\infty} xN(x)dx = \frac{1}{\sqrt{2\pi\sigma^2}} \int_{-\infty}^{+\infty} xe^{-(x-\mu)^2/2\sigma^2} dx$$

Let y = x-μ, $dx = dy$, and then, from Appendix 15.6,

$$\frac{1}{\sqrt{2\pi\sigma^2}} \int_{-\infty}^{+\infty} (y+\mu)e^{-\frac{y^2}{2\sigma^2}} dy = 0 + \frac{1}{\sqrt{2\pi\sigma^2}} \mu(\sqrt{2\pi\sigma^2}) = \mu,$$

as is expected.

Kalman Filter

To calculate the variance of the Gaussian distribution we evaluate

$$Var(x) = E[(x - \mu)^2] =$$

$$\frac{1}{\sqrt{2\pi\sigma^2}} \int_{-\infty}^{+\infty} (x - \mu)^2 \exp{-\frac{1}{2\sigma^2}(x - \mu)^2} dx$$

Let $z = (x - \mu)/\sigma$ and $\sigma dz = dx$ and then

$$Var(x) = \frac{1}{\sqrt{2\pi\sigma^2}} \int_{-\infty}^{+\infty} \sigma^2 z^2 \exp{-\frac{1}{2}z^2} \sigma dz$$

$$= \frac{\sigma^2}{\sqrt{2\pi}} \int_{-\infty}^{+\infty} z^2 \exp{-\frac{1}{2}z^2} dz$$

Using the result from Appendix 15.6 we obtain

$$Var(x) = \frac{\sigma^2}{\sqrt{2\pi}} 2\frac{1}{2}\sqrt{2\pi} = \sigma^2,$$

as is expected.

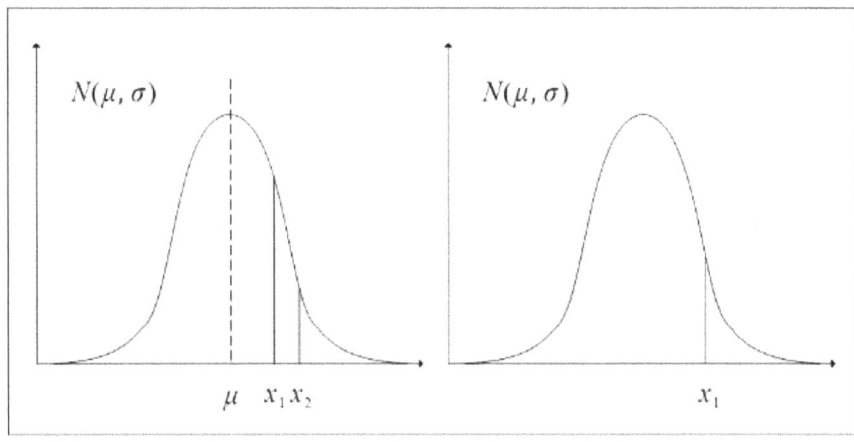

Figure 9 – Gaussian Probability Distribution

4.9.1 Body and Tail of a Distribution

The body of a distribution refers to the centre of the distribution. This is where most of the probability is found. The tails of the

Kalman Filter

distribution are the extreme upper and lower parts of the curve where the probability of occurrence is very low.

4.10 Multivariate Gaussian Distribution

The one-dimensional (scalar) Gaussian distribution for a variable x with mean and standard deviation μ, σ is given by

$$P(x|\mu, \sigma) = \sqrt{\frac{1}{2\pi\sigma^2}} exp[-(x-\mu)^2/2\sigma^2]$$

The natural generalisation for a vector valued random variable is given by

$$P(x|m, R) = \frac{1}{Z(R)} exp\left[-\frac{1}{2}(x-m)^T R(x-m)\right]$$

where:

$$Z(R) = \sqrt{\det(R/2\pi)}^{-1}$$

and R is the inverse of the covariance matrix.

4.11 Random Vector

A random variable is characterised by its probability density function. The aggregate behaviour of many samples can be determined from various measures. First is the mean, or expected value. The mean of a random variable x is denoted by $\bar{x}$ or $E(x)$. E is the linear expectation operator. The mean alone does not characterise the random variable very well. The spread about the mean is measured by the variance. The variance is the mean of the values $(x - \bar{x})^2$. The variance is denoted by

$$\sigma^2 = E\{(x-\bar{x})^2\}$$

The square root of the variance is called the standard deviation (σ). It has the merit of having the same dimensions as the mean and makes the comparison between the mean and the spread more meaningful.

The measure of dependence between two random variables x and y is given by the covariance, defined by

Kalman Filter

$$p(x, y) = E\{(x - \bar{x})(y - \bar{y})\}$$

When x and y are the same random variable, the covariance reduces to the variance.

The correlation coefficient of x and y is given by

$$\rho(x, y) = \frac{p(x,y)}{\sigma_x \sigma_y}$$

The correlation coefficient lies between the values ± 1. If the value of ρ is 1 then the values are highly correlated. If it is 0, then they are uncorrelated. If it is -1 then they are anti-correlated.

A collection of random variables $x_1, x_2, \ldots x_n$ can be collected into an array and this is a random vector.

$$x = \begin{pmatrix} x_1 \\ x_2 \\ \ldots \\ x_n \end{pmatrix}$$

The mean vector is the vector of the mean values. The covariance matrix is given by the product of the column vector and the row vector (to result in a matrix) as follows

$$P = E\{(x - \bar{x})(y - \bar{y})^T\}$$

The ij element is the covariance $p(x_i, x_j)$ and the ii diagonal elements are the variance $\sigma^2(x_i)$. The covariance matrix is symmetric and is positive semi-definite.

4.12 Random Process

Systems often include random or unknown components. In such cases the system is often best described probabilistically. Such processes are called random processes. Examples are systems where the measurements have, superimposed on the signal, random noise. In this case, the random noise could be a voltage fluctuation, such as Johnson noise (thermal), shot noise (from discrete charges) or 1/f noise (a type of low frequency noise found in nature).

Kalman Filter

The term 'stationary' refers to the time invariance of various aspects of the statistics of a random process. Stationary noise, such as thermal noise, will appear the same, statistically, whether it is observed today or tomorrow (under the same conditions).

4.12.1 Strict Sense Stationary

A random process is said to be SSS if the distribution depends only on time differences and not on the absolute time.

$$p(x_t|x_{t_1}, x_{t_2}, \ldots) = p(x_t|x_{t-\tau_1}, x_{t-\tau_2}, \ldots)$$

4.12.2 Wide Sense Stationary

A random process is said to be WSS (a weaker condition than SSS) if the means and covariances of the process are independent of time. Namely, for the mean and autocorrelation,

$$<X(t)> = \mu \quad (a\ constant, independent\ of\ t)$$
$$R_X(t_1, t_2) = R_X(|t_1 - t_2|)$$
$$(a\ function\ only\ of\ the\ time\ difference)$$

A SSS system is also WSS.

4.12.3 White Noise

White noise is a purely random input. It is stationary random signal but it has equal intensity at all frequencies. Statistically it is a signal whose samples are unrelated (uncorrelated) zero mean random variables. Often each sample is considered to come from a Gaussian distribution with zero mean. This is referred to as Additive White Gaussian Noise.

Statistically, a discrete white noise process W(k) has the properties

$$<W(k)> = 0 \quad \quad zero\ mean$$
$$<W(k)W(k+n)> = \sigma^2 \delta(n) \quad uncorrelated\ samples$$

Gaussian white noise is stationary and ergodic.

4.12.4 The Statistical Ensemble

A statistical ensemble (or ensemble) is an assembly of N similar systems. The number of systems is envisaged to be arbitrarily large ($N \to \infty$). Each system is imagined to have been prepared in the same way and to be subject to the same set of observations. If the outcomes of the systems can be enumerated then, if the number of occurrences of a particular outcome $'i'$ is labelled N_i, the probability of occurrence of that outcome is given by the fraction

$$P_i = \frac{N_i}{N} \text{ as } N \to \infty$$

Thus, the probability of an outcome can be compared with experiment.

For any single system, the observation can be taken over time, rather than over the ensemble. In some cases, the ensemble average and the time average are identical.

A statistical ensemble is illustrated in Figure 10.

The statistical ensemble represents a probability distribution of the states of the system and this provides a means to visualise the idea of the probability of the system being in a particular state.

Kalman Filter

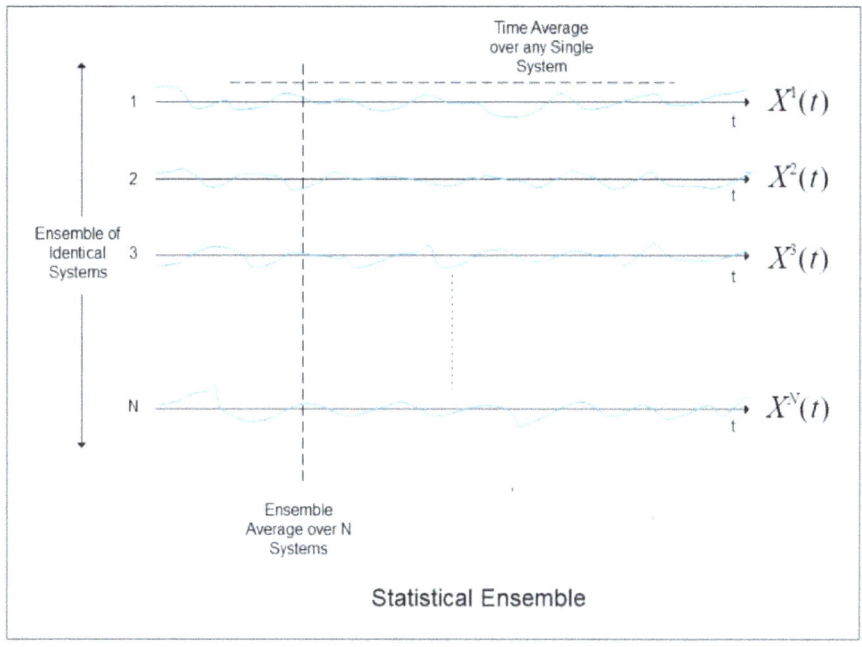

Figure 10 – Statistical Ensemble

4.12.5 Ergodicity

A system is ergodic if the average output of an ensemble of such systems, at any given time, is the same as the average outcome of any single system over time. A common example can be found in coin flipping. The statistical outcome is unchanged if 1000 people toss a coin once or if one person tosses a coin 1000 times. The mean and variance of the outcome are the same in either case. Ergodicity is an important concept because it provides the means to study long term average behaviour of systems from a knowledge of ensemble averages. The inference of a general statement from a particular experiment assumes ergodicity.

Ergodicity states that an ensemble average, at any particular time, is equal to a time average of a single instance of the system, and so an ergodic process is also stationary.

Kalman Filter

4.12.6 Example: Mean Ergodicity

A system is said to be mean ergodic if

$$\lim_{t \to \infty} E[(X(t) - \mu)^2] = 0$$

This is the same as

$$\lim_{t \to \infty} Var[X(t)] = 0$$

From the expression

$$E(\bar{X}^2(t)) = E\left[\left(\frac{1}{t}\int_0^t X(\tau)d\tau\right)^2\right] =$$
$$E\left[\frac{1}{t^2}\int_0^t X(\tau_1)d\tau_1 \int_0^t X(\tau_2)d\tau_2\right]$$

$$E(\bar{X}^2(t)) = E\left[\frac{1}{t^2}\int_0^t d\tau_1 \int_0^t X(\tau_1)X(\tau_2)d\tau_2\right]$$

$$E(\bar{X}^2(t)) = \frac{1}{t^2}\int_0^t d\tau_1 \int_0^t d\tau_2 R(\tau_1, \tau_2)$$

For a wide sense stationary process, the autocorrelation is a function only of the time difference,

$$E(\bar{X}^2(t)) = \frac{1}{t^2}\int_0^t d\tau_1 \int_0^t d\tau_2 R(\tau_1 - \tau_2)$$

this reduces to the integral given by

$$E(\bar{X}^2(t)) = \frac{2}{t^2}\int_0^t d\tau(t - \tau)R(\tau)$$

This integral is derived in appendix 15.4.

4.13 Bayes Theorem

Bayes rules is used to determine the probability of an event which is based upon the knowledge of prior related events. For two events A and B, that occur with marginal probabilities P(A) and P(B) respectively, the joint probability that both events occur is P(A, B). The conditional probability that event A occurs given that event B has occurred is P(A|B). Similarly, the conditional probability that event B occurs given that even A has occurred is given by P(B|A). It should be clear that, by symmetry, the joint probability is given by

Kalman Filter

$$P(A,B) = P(A|B)P(B) = P(B|A)P(A)$$

and hence

$$P(A|B) = \frac{P(B|A)P(A)}{P(B)}$$

This is Bayes theorem. There is a nomenclature associated with Bayes theorem and this is summarised in Figure 11.

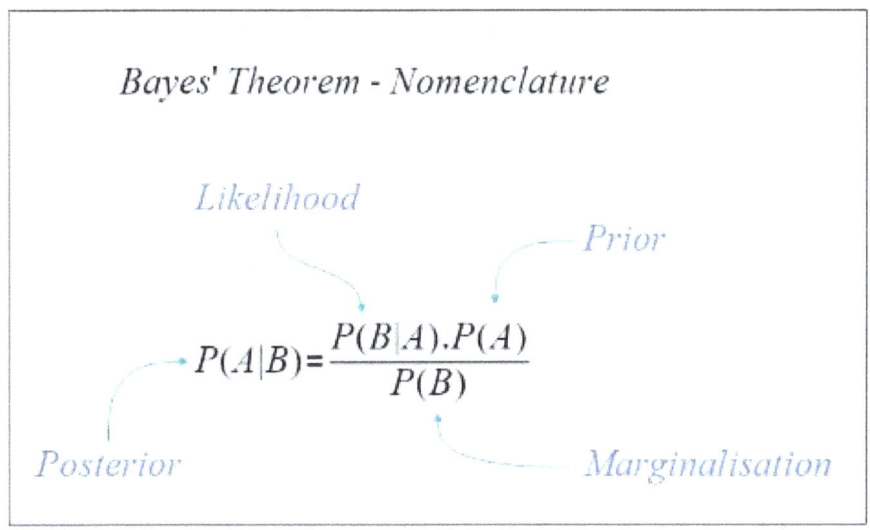

Figure 11 – Bayes Theorem Nomenclature

The terms are defined as follows:

Prior: The probability that A is true – prior information.

Likelihood: The probability that B is true given that A is true.

Marginalisation (Evidence): The probability that B is true.

Posterior: The probability that A is true given that B is true – after evidence is accounted for.

4.13.1 Example: Balls Drawn from a Bag

Bag A contain 4 white balls and 6 red balls.

Bag B contains 4 white balls and 3 red balls.

Kalman Filter

One ball is drawn at random from an unknown bag. The ball is red.

What is the probability that the ball was drawn from Bag A?

The probability of choosing bag A is 0.5, as is the probability of choosing bag B. Hence,

$$P(A) = P(B) = 0.5$$

The probability of drawing a red ball from Bag A is

$$P(R|A) = \frac{6}{10}$$

The probability of drawing a red ball from bag B is

$$P(R|B) = \frac{3}{7}$$

Hence, the probability of drawing from bag A, a red ball, is

$$P(A|R) = \frac{P(R|A)P(A)}{P(R)} = \frac{P(R|A)P(A)}{P(R|A)P(A)+P(R|B)P(B)}$$

$$= \frac{\left(\frac{3}{5}\right)\left(\frac{1}{2}\right)}{\left(\frac{1}{2}\right)\left(\frac{3}{5}\right)+\left(\frac{1}{2}\right)\left(\frac{3}{7}\right)} = \frac{7}{12}$$

4.14 The Law of Total Probability

The law of total probability is useful when the probability of an event is unknown but it occurs in a number of different scenarios and the probability of each separate scenario is known.

In terms of conditional probabilities, if we have an event A that can occur via a number of other events, $B_1,...,B_n$ then in each case, if we know the conditional probabilities $P(A|B_i)$ and the marginal probability of each of the B events, then the probability of event A is given by

$$P(A) = \sum_{i=1}^{n} P(A|B_i)P(B_i)$$

The events B_i are non-overlapping. The B_i must sum to one because they are a partition of all possible outcomes.

Kalman Filter

4.14.1 Example: Law of Total Probability

Draw two cards from a shuffled deck. The cards are replaced after being drawn. What is the probability that that the second card is an Ace?

Let A_1 be the event the first card is an Ace.

Let Z be the event that the first card is not an Ace.

Let A_2 be the event that the second card is an Ace.

Note that A_1 and Z form a disjoint partition the first event.

Then, using an obvious notation,

$$P(A_2) = P(A_2|A_1)P(A_1) + P(A_2|Z)P(Z)$$

4.15 Likelihood

Colloquially the terms probability and likelihood appear to be similar but in statistics there is a difference between the terms. This is most easily explained using an example. First, in summary, note that probability applies to possible results whereas likelihood applies to hypotheses. We can see this as follows.

Consider a distribution of measured values, with mean μ and standard deviation σ. First, we have a measurement of value x = X_0. The probability that we measure a value between x_1 and x_2, for this distribution, is given by the area under the curve between the values x_1 and x_2.

$$p(x_1 < x < x_2 \,|\mu, \sigma).$$

On the other hand, if we have observed a value x_1. The likelihood of the distribution having mean μ and standard devaition σ, given the measurement value x_1, is given by the value of the probability curve corresponding to the value x_1. This is given by

$$L(\mu, \sigma | x = x_1)$$

So, we have, in summary:

The probability of a data measurement given a fixed distribution.

Kalman Filter

The likelihood a distribution given observed data. Figure 12 shows the likelihoods, L_1 and L_2, of observing a value x_1, given two different distributions.

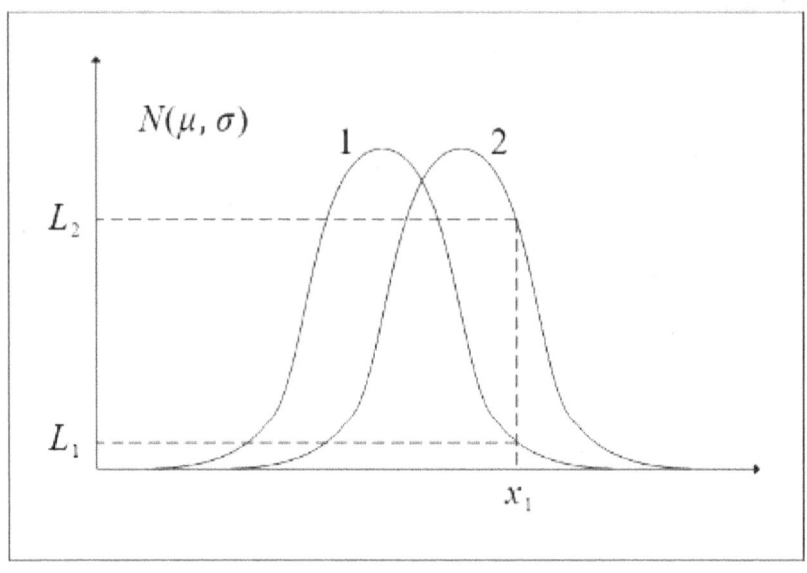

Figure 12 – likelihood

4.15.1 Maximum Likelihood Estimator

Given a set of observations, the data corresponds to a particular probability distribution, such as Gaussian, Exponential, etc. The general appearance of the data may favour a particular distribution. For example, if the data is clustered about the mean, and if it is also symmetric about the mean, a Gaussian distribution seems likely. Once a candidate distribution has been identified the next step is to determine the parameters of the distribution. For the Gaussian this requires an estimate of the mean and the standard deviation.

For a number of points along the x-axis it is possible to calculate the likelihood of observing the given data. The point at which the likelihood is maximised is then chosen as the mean of the distribution. Similarly, for the standard deviation, we plot the

Kalman Filter

likelihood of observing the data for different values of the standard deviation. We then choose the value that maximises the likelihood of observing the data. Thus, we have the maximum likelihood estimate of the mean and standard deviation of the distribution.

4.15.2 Example: Gaussian

Once a candidate PDF has been identified then it is possible to estimate the mean value from the observed data. The PDF is parameterised in terms of the unknown quantity and we denote the unknown quantity, the mean, as θ. The data observations are assumed to be uncorrelated and with identical means and variances. Then we consider a function of θ and σ which gives the PDF of the observed data $\{x_x, x_2, \ldots, x_N\}$ for any value of θ.

$$p(x|\theta, \sigma) = \frac{1}{\sqrt{2\pi\sigma^2}} exp\left[-\frac{(x_1-\theta)^2}{2\sigma^2}\right] \frac{1}{\sqrt{2\pi\sigma^2}} exp\left[-\frac{(x_2-\theta)^2}{2\sigma^2}\right] \ldots$$

$$\ldots \frac{1}{\sqrt{2\pi\sigma^2}} exp[-(x_N - \theta)^2/2\sigma^2]$$

$$p(x|\theta, \sigma) = \left(\frac{1}{\sqrt{2\pi\sigma^2}}\right)^N exp\left[-\frac{1}{2}\sum_{i=1}^{N}(x_i - \theta)^2/\sigma^2\right]$$

This gives the PDF of a set of data $x = \{x_i\}$ for any value of θ. The probability that an observed value of x is in the range $[a, b]$ is

$$P(a \le x \le b) = \int_a^b p(x|\theta, \sigma) dx$$

This is a measure of the likelihood because when p is large, we expect to observe many values of x in that region. The function of θ, $p(x|\theta, \sigma)$, is called the likelihood function. To estimate the mean we maximise the likelihood function wrt to θ. This is the maximum likelihood estimate, denoted by $\hat{\theta}_{ML}$.

It is convenient to deal with the exponential by using the chain rule, to yield, for a function f,

$$\frac{d}{dx} \log f = \frac{1}{f}\frac{df}{dx}$$

Hence the derivative of $\log f = 0$ if $\frac{df}{dx} = 0$. We can proceed with

Kalman Filter

the log of f rather than the function f. Hence, for convenience, defining the 'log' term as

$$-\log p(x|\theta, \sigma) \equiv \ell(x|\theta, \sigma)$$

we can write the exponential as

$$\ell(x|\theta, \sigma) = \frac{N}{2}\log(2\pi\sigma^2) + \frac{1}{2}\sum_{i=1}^{N}(x_i - \theta)^2/\sigma^2$$

and then, to find the maximum value of the variable θ, differentiate

$$\frac{d\ell}{d\theta} = -\frac{1}{2\sigma^2}\sum_{i=1}^{N} 2(x_i - \theta) = 0$$

$$= -\frac{1}{\sigma^2}[(\sum_{i=1}^{N} x_i) - N\theta] = 0$$

$$\sum_{i=1}^{N} x_i - N\hat{\theta}_{ML} = 0$$

Hence

$$\hat{\theta}_{ML} = \frac{1}{N}\sum_{i=1}^{N} x_i$$

and therefore, for the Gaussian distribution, the maximum likelihood estimate of the distribution mean is the sample mean.

Similarly, to find the maximum likelihood estimate of the variance, differentiate wrt σ^2 to give

$$\frac{\partial \ell}{\partial \sigma^2} = \frac{N}{\sigma^2} - \frac{1}{\sigma^4}\sum_{n=1}^{N}(x_n - \mu)^2$$

Setting this to zero gives

$$N - \frac{1}{\sigma^2}\sum_{n=1}^{N}(x_n - \mu)^2 = 0$$

$$\widehat{\sigma^2}_{ML} = \frac{1}{N}\sum_{n=1}^{N}(x_n - \mu)^2$$

and, as for the mean, the maximum likelihood estimate of the distribution variance is the sample variance.

The intuitive forms for the mean and standard deviation of the distribution are seen to be the maximum likelihood estimates. This puts both concepts on a solid foundation.

Kalman Filter

5 Least Squares Methods

A common technique, which we will use, is based on the least squares fit. In this example we consider a set of points which represent measurements made of a function which we expect to be linear (a straight line). The aim is to find the best line that fits the data according to some criteria. This a called linear regression.

The straight line is determined by two parameters, m the gradient, and c the y-intercept. We must find these values in terms of the data points. The required line is given by

$$y = mx + c$$

and the data points are given by the set of x-y pairs $\{i, y_i\}$ where i, the x-coordinate, runs from -M to +M.

The quality of the fit is measured by the sum of the squared distances from each y point to the line (denoted by S). In order to avoid cancelations, the lengths are squared and then summed to give an overall measure of the goodness of the fit. To find the best line, according to our criteria, we minimise the measure with respect to m and c.

$$S = \sum_{i=-M}^{+M}(y(i) - y_i)^2 = \sum_{i=-M}^{+M}(mi + c - y_i)^2$$

Minimise with respect to m and c to give

$$\frac{\partial S}{\partial m} = 2\sum_{i=-M}^{+M}(mi + c - y_i)i = 0$$

$$\frac{\partial S}{\partial c} = 2\sum_{i=-M}^{+M}(mi + c - y_i) = 0$$

The first equation expands to give

$$\sum_{i=-M}^{+m} i^2 m + c\sum_{i=-M}^{+M} i - \sum_{i=-M}^{+M} y_i i = 0$$

The second term is zero because $\sum_{i=-M}^{+M} i = 0$. Hence, the slope m is given by a weighted sum of the data values

$$m = \frac{\sum_{i=-M}^{+M} i\, y_i}{\sum_{i=-M}^{+M} i^2}$$

From the second equation, the value of c, the y-intercept, is the

Kalman Filter

average of the data values.

$$c = \frac{\sum_{i=-M}^{+M} y_i}{2M+1}$$

This approach can be used for fitting other curves, such as higher order polynomials, to a set of data points.

For our purposes it is the illustration of the method that is of interest because it can be applied to more general problems. The principle is used in the derivation of the Kalman filter equations.

6 Digital Filter

The digital filter is a useful introduction to the ideas behind a Kalman filter. The digital filter processes an input stream of digital data to produce a delayed output stream of digital data. The output is filtered in the sense that it may, for example, contain only certain frequency components of the original data. Another use is to remove unwanted random noise. The concept is illustrated in Figure 13. The filter in the diagram has removed undesirable random noise from the signal.

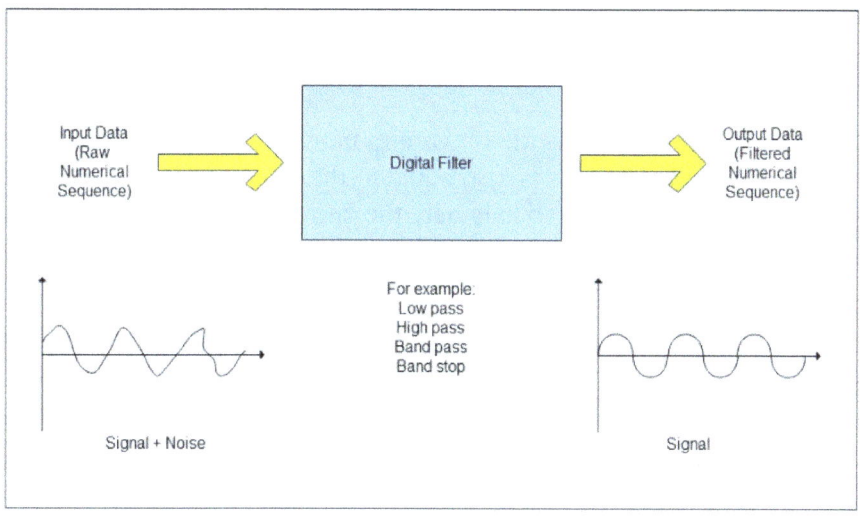

Figure 13 – Digital Filter

Digital filters are programmable, and can be implemented on a microprocessor, are easily changed, are stable wrt to parameter aging, and can be designed to be adaptive.

A digital filter operates on an input data stream $(x_0, x_1, ...)$ to produce an output stream $(y_0, y_1, ...)$. There are two main types of digital filter: recursive (IIR) and non-recursive (FIR).

The digital filter concept spans both DSP and statistical time series analysis. There are different terms for these filters in the different disciples. The terminology is summarised in Table 6-1.

Kalman Filter

Filter Type	Time Series Analysis	DSP
Non-Recursive	Moving Average (MA)	FIR
Recursive	Autoregressive (AR)	IIR

Table 6-1 – Filter Nomenclature

In time series analysis, the general term for a model of this kind is $ARMA(M, N)$, where M and N define the order of combined AR and MA filters respectively. An ARMA model consists of both AR and MA parts combined into a single model. An FIR filter is illustrated in Figure 14. The IIR filter is illustrated in in Figure 15.

The order of a filter is the maximum number of delay elements used in the filter. This is also equal to the number of previous inputs that are stored to calculate the current output.

6.1 Non-Recursive Filter

The FIR filter is illustrated in Figure 14. The defining equation is

$$y[n] = a_0 x[n] + a_1 x[n-1] + \cdots + a_q x[n-q]$$

where the output at time n is defined by a weighted sum of a finite number of previous inputs. The filter coefficients may be positive or negative. This is a FIR filter of order N.

Kalman Filter

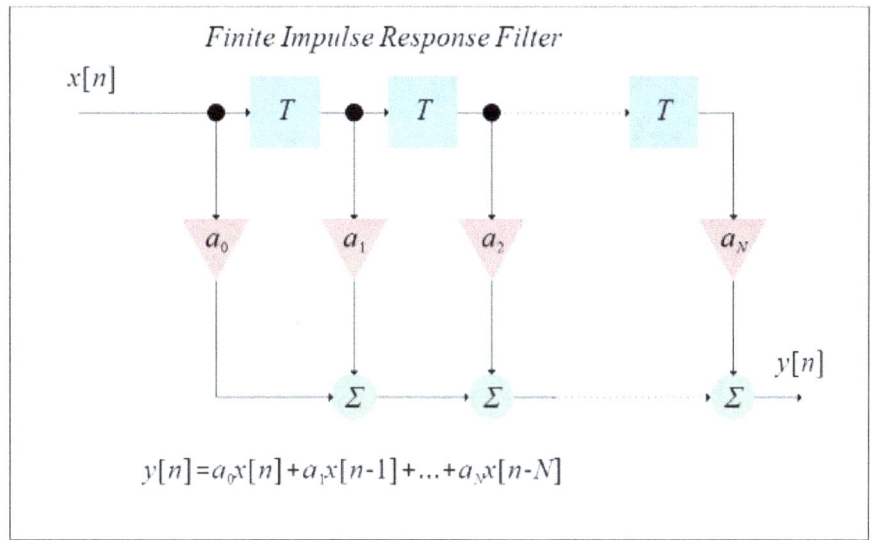

Figure 14 – FIR Filter

This kind of filter resembles an $\alpha - \beta$ filter as discussed in section 8.

6.2 Recursive Filter

The IIR filter is illustrated in Figure 15. The defining equation for this filter is

$$y[n] =$$
$$a_0 x[n] + a_1 x[n-1] + a_2 x[n-2] + a_3 x[n-3] -$$
$$b_1 y[n-1] - b_2 y[n-2] - a_3 y[n-3]$$

where the output at time n is defined by a weighted sum of a finite number of previous inputs and outputs. The filter coefficients may be positive or negative. This is a IIR filter of order 3.

This kind of filter resembles a Kalman filter as discussed in 11.

Kalman Filter

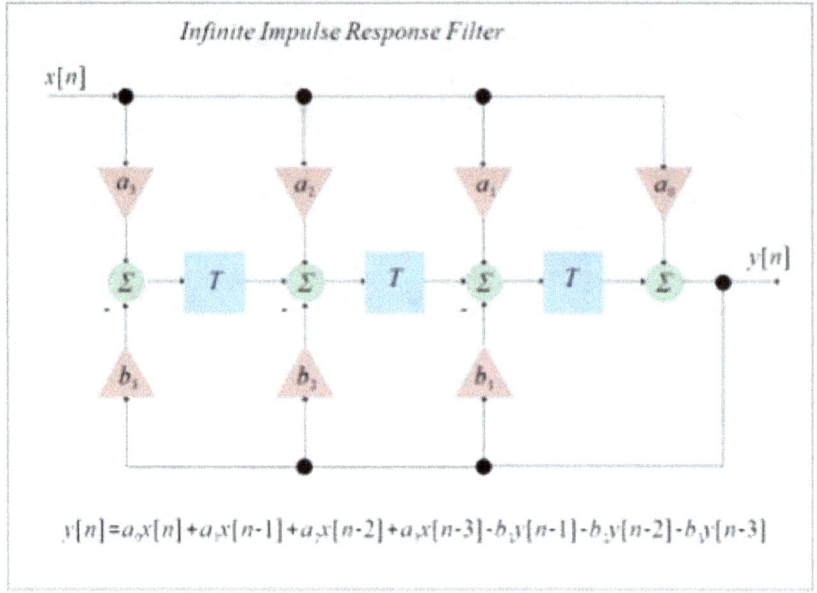

Figure 15 – IIR Filter

6.3 Stochastic Processes

A stochastic (random) process is any process describing the evolution in time of a random phenomenon. A stochastic process x(t) can be described in terms of the probability of observing a sample value x at a time t, given the system history. This is expressed as a conditional probability distribution

$$p(x(t_k)|x(t_{k-1}), \ldots, x(t_{k-N}))$$

If t is a real number, then x(t) is a continuous-time process. If t is an integer, then x(t) is a discrete-time process. Similarly, we have continuous-state and discrete-state processes corresponding to the cases where the values of x are either continuous or discrete, respectively.

If the following condition is true

$$p(x(t_k)|\, x(t_{k-1}), \ldots, x(t_{k-N})) = p(x(t_k)|\, x(t_{k-1}))$$

then the process is called a Markov process.

Kalman Filter

7 Markov Processes

Before discussing in detail Markov processes and associated concepts, it is helpful to define a number of terms for easy reference:

Markov Process

If the conditional probability distribution of a stochastic process is limited to a finite history of N terms

$$p(x_t|x_{t-1}, x_{t-2}, \ldots, x_{t-N})$$

then the process is said to be an Nth order Markov process.

If the conditional probability distribution depends only on the previous value

$$p(x_t|x_{t-1})$$

then it is called a Markov process.

Markov Chain

If a Markov process has a finite number of states, then it is called a Markov chain. A Markov chain can be continuous-time or discrete-time depending on whether the t value is continuous or discrete.

Hidden Markov Model

A Hidden Markov Model (HMM) is a statistical model in which the system being modelled is assumed to be a Markov process with hidden (unobservable) states. The hidden states are a Markov process whose values determine, probabilistically, the observed states, whose known values are referred to as observations.

7.1 Markov Chain

A Markov chain consists of the following components:

    A state space:

$$\{S_1, \ldots, S_N\}$$

Kalman Filter

The Markov assumption:

$$P(e_t|e_{t-1}, e_{t-2}, \ldots) = P\{e_t|e_{t-1}\}$$

A Transition Matrix consisting of the probabilities of transitions between states.

An example of such a system, a state diagram with three states, is illustrated in Figure 16. The corresponding transition matrix is presented in Table 7-1. Note that the rows of the transition matrix must sum to 1.0 to conserve probability because the destination states are exhaustive; the total of the transition probabilities from any given state to all other states must be one.

A Markov chain can reach a steady state. Let the probability of being in state i in the steady state be π_i for i in the range 1, 2, 3. The sum of the probabilities must always be equal to 1 and so

$$\pi_1 + \pi_2 + \pi_3 = 1$$

In the steady state, application of the transition matrix results in no change to the state average occupancy. Hence, for a steady state we require the following conditions. The total probability of entering state 1 from state 1, state 2, and state 3 is given by:

$$0.3\,\pi_1 + 0.4\,\pi_2 + 0.2\,\pi_3 = \pi_1$$

For state 2, similarly,

$$0.6\,\pi_1 + 0.2\,\pi_2 + 0.3\,\pi_3 = \pi_2$$

Substitute for $\pi_3 = 1 - \pi_1 - \pi_2$ and solve the simultaneous equations to give:

$$\pi_1 = \frac{28}{3(31)},\ \pi_2 = \frac{11}{31},\ \pi_3 = \frac{32}{3(31)}$$

Kalman Filter

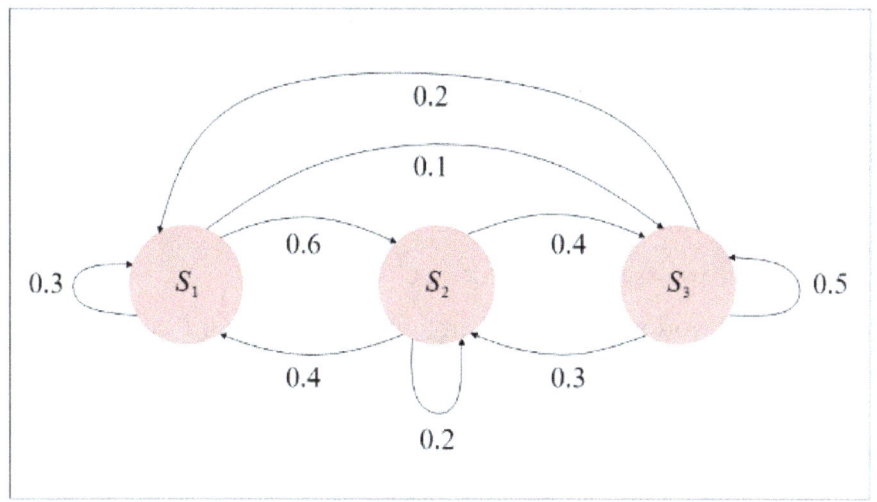

Figure 16 – Markov Chain

		State (t = n+1)			
		State 1	State 2	State 3	Total
State (t = n)	State 1	0.3	0.6	0.1	1.0
	State 2	0.4	0.2	0.4	1.0
	State 3	0.2	0.3	0.5	1.0

Table 7-1 – Transition Matrix

7.2 Stochastic Matrix

A stochastic matrix (transition matrix) is a matrix that describes the transitions of a Markov chain. It is a square matrix whose entries describe the probabilities of transitions, as seen above. Because of the probabilistic nature of the matrix, all the entries must be non-negative (a probability) and each row must sum to one (the sum of probabilities of all possible events must be one).

Kalman Filter

This can be expressed as follows. The state transition coefficients are defined to be, for ($state\ i \rightarrow state\ j$),

$$a_{ij} = P(s_{n+1} = \alpha_j | s_n = \alpha_i)$$

and the constraint of conservation of probability is

$$\sum_{j=1}^{N} a_{ij} = 1$$

This is expressed by a matrix such as Table 7-1.

In other words, if we are in any given state then the next state must be one of the possible states and so the sum over the probabilities of all possible states must be one; the probabilities of leaving a state must sum to one.

7.3 Hidden Markov Model

A Hidden Markov Model is a probabilistic model where there are both observed states and hidden (internal states). The hidden states are involved in the generation of the observed states. It is possible to learn about the hidden states from information obtained from the observed states.

A Hidden Markov Model is illustrated in Figure 17. There are two hidden states and three observed states. There are two important tables to note: The Transition Probabilities and the Emission Probabilities. These are illustrated in Table 7-2 and Table 7-3.

Kalman Filter

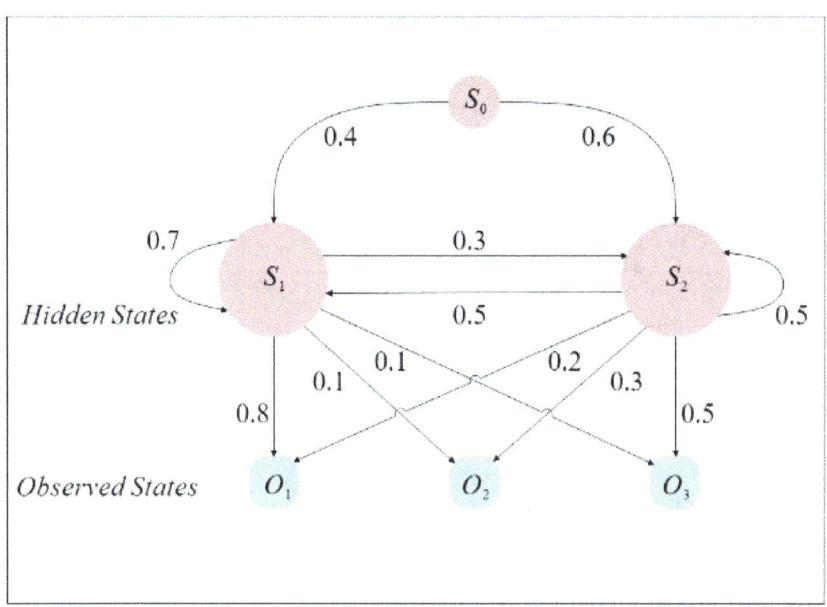

Figure 17 – Hidden Markov Model

Transition Probabilities

	S1(t)	S2(t)	Total
S1(t-1)	0.8	0.3	1.0
S2(t-1)	0.5	0.5	1.0

Table 7-2 – Transition Probabilities

Kalman Filter

Emission Probabilities

	O1	O2	O3	Total
S1	0.8	0.1	0.1	1.0
S2	0.2	0.3	0.5	1.0

Table 7-3 – Emission Probabilities

The hidden state determines the probabilities that the observed states will be occupied. Given a set of observations we wish to estimate the most likely hidden state.

The hidden states affect the observed states. If we have three observations x1, x2, and x3, at times 1, 2, and 3, then we require the most likely set of hidden states that resulted in the observations. This is given by

$$\frac{MAX}{h_1, h_2, h_3} P(o_1 = x1, o_2 = x2, o_3 = x3, S = \{h_1, h_2, h_3\})$$

We require the combination of the three hidden variables h_i that maximise the probability.

This probability is given by

$$P(x_3|x_2, x_1, h_1, h_2, h_3)P(x_2|x_1, h_1, h_2, h_3)$$
$$P(x_1|h_1, h_2, h_3)P(h_3|h_2, h_1)P(h_2|h_1)P(h_1)$$

For the first three conditional probabilities the observation on any given time depends only on the hidden state at that time. Then, in addition, applying the Markov condition to the second three probabilities gives

$$P(x_3|h_3)P(x_2|h_2)P(x_1|h_1)P(h_3|h_2)P(h_2|h_1)P(h_1|S_0)$$

where the first hidden state h_1 depends only upon the start state

Kalman Filter

S_0.

Using the Transmission and Emission probability tables, this expression can be maximised, by plugging in the values for combinations of hidden states, to determine the most probable sequence of hidden states, given the observations.

7.3.1 The Kalman Filter Viewed as a HMM

The Kalman filter can be thought of as a Hidden Markov Model. The measurements define the observed states and the internal dynamics are defined via a state vector which contains the hidden states, to be determined. In the Kalman filter, in contrast to the HMM, the state of the system is determined without a knowledge of the probability distribution of the hidden states.

7.4 Chapman-Kolmogorov Equations

Using the Markov property and the law of total probability, the Chapman-Kolmogorov (C-K) equations are derived. These equations describe the transition from state i to state j in $n + m$ steps, with an intermediate stop at state k after n steps. A sum is taken over all possible intermediate k values.

Let

$$P_{ij}^n = P(X_{n+m} = j | X_m = i), \quad n, m =\geq 0$$

Then P_{ij}^n is the probability of moving from state i to state j in n steps. The matrix

$$P^{(n)} = \begin{pmatrix} p_{00}^n & p_{01}^n & \cdots \\ p_{10}^n & p_{11}^n & \cdots \\ \cdots & \cdots & \cdots \end{pmatrix}$$

is called the n-step transition probability matrix. $P^{(0)}$ is the identity matrix and $P^{(1)} = P$ is the transition probability matrix of the Markov chain.

For a transition from a state i to a state j in $n + m$ steps, a

Kalman Filter

Markov chain must enter some state k, after n transitions, and then move from k to j in m transitions. From the Law of Total Probability, for a set of mutually exclusive events

$\{X_n = k\}, k \geq 0$

$$P_{ij}^{n+m} = P(X_{n+m} = j \mid X_0 = i)$$
$$= \sum_{k=0}^{\infty} P(X_{n+m} = j \mid X_n = k, X_0 = i) P(X_n = k \mid X_0 = i)$$
$$= \sum_{k=0}^{\infty} P(X_{n+m} \mid X_n = k) P(X_n = k \mid P_0 = i)$$

Markov property

$$= \sum_{k=0}^{\infty} p_{kj}^m p_{ik}^n = \sum_{k=0}^{\infty} p_{ik}^n p_{kj}^m$$

In matrix form this is

$$P^{(n+m)} = P^{(n)} P^{(m)}$$

These are the Chapman-Kolmogorov equations.

In general

$$P^{(n)} = P^n.$$

7.4.1 Example: C-K Equations

A Markov chain with two states and associated transition probabilities is shown in Figure 18. Transition probabilities for times 0 to 1 are given by the matrix

$$P = \begin{pmatrix} P_{AA} & P_{AB} \\ P_{BA} & P_{BB} \end{pmatrix}$$

where P_{ij} is the probability of transition between states $i \to j$.

Kalman Filter

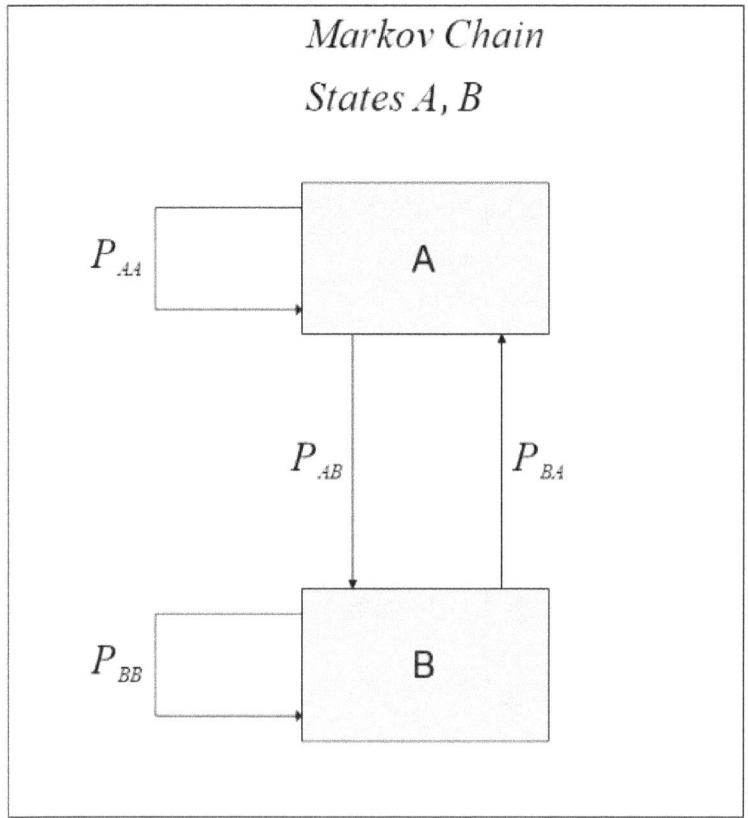

Figure 18 – CK Equation Example

The following probabilities are given as shown.

A direct transition from A to B:

$$P(B \text{ at } t = 1 | A \text{ at } t = 0) = P_{AB}$$

The sum of $A \to A$ followed by $B \to A$ and $B \to A$ followed by $B \to B$.

$$P(A \text{ at } t = 2 | B \text{ at } t = 0) = P_{BA}P_{AA} + P_{BB}P_{BA}$$

This is the entry BA of the matrix product PP. Note that it is summed over both A and B intermediate states. As per the CK

Kalman Filter

equation, we can obtain this by matrix multiplication.

For the case
$$P(B \text{ at } t = 2 | A \text{ at } t = 0)$$
we proceed by matrix multiplication.
$$PPP = \begin{pmatrix} N/A & P3_{AB} \\ N/A & N/A \end{pmatrix}$$
where:
$$P3_{AB} =$$
$$P_{AA}P_{AA}P_{AB} + P_{AA}P_{AB}P_{BB} + P_{AB}P_{BA}P_{AB} + P_{AB}P_{BB}P_{BB}$$
The matrix multiplication takes care of the sum over all intermediate states
$$A \to A, B \to A, A \to B, B \to B.$$

8 Alpha-Beta Filter

The $\alpha - \beta$ filter is a simple filter that combines observation and estimation. It provides an introduction to the principles that underlie the Kalman filter. The terms α and β are two constant scaling factors that are used to weight the measurements. The coefficient α corresponds to the weighting for the measurements and the β coefficient corresponds to the weighting of the change in measurement. The filter is described by using the following example, as in Figure 19.

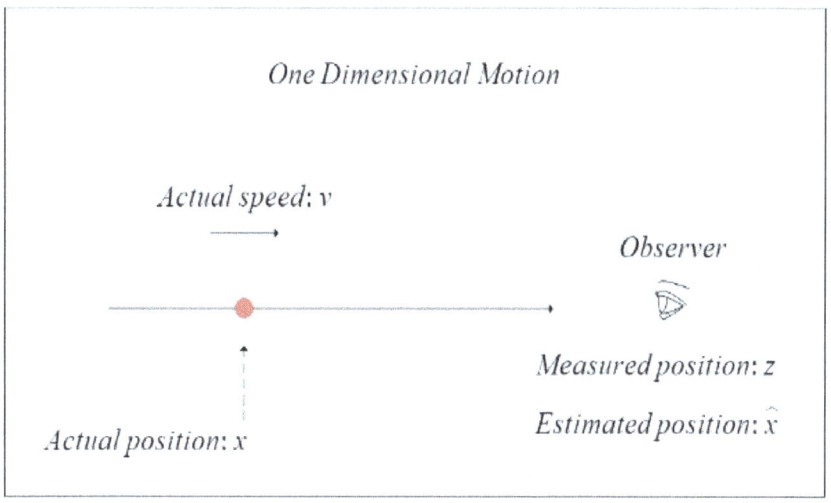

Figure 19 – Moving Object

A one-dimensional system, consisting of a moving object with uniform speed is to be observed. The position is denoted by x. The position measurement is denoted by z. Hence, the speed of the object is given by

$$v = \frac{dx}{dt}$$

The target is tracked by a sensor, such as radar, at regular intervals denoted by T. Given the assumptions above the system

Kalman Filter

can be modelled as follows (where k denotes time increments)

$$x(k+1) = x(k) + T\dot{x}(k) \quad \text{linear motion}$$
$$\dot{x}(k+1) = \dot{x}(k) \quad \text{constant speed}$$

These equations extrapolate the state; the next state is computed from the current state. The estimated state of the system (position) is denoted by $\hat{x}$ and the state estimate is given by the position estimate and the speed estimate:

$$\hat{x}(k|k) = \hat{x}(k|k-1) + \alpha(z(k) - \hat{x}(k|k-1))$$
$$\hat{\dot{x}}(k|k) = \hat{\dot{x}}(k|k-1) + \beta\left(\frac{z(k) - \hat{x}(k|k-1)}{T}\right)$$

where:

$\hat{x}(k|k-1)$ is the predicted position at time k given the measurement at time k – 1.

$\hat{x}(k|k)$ is the estimated position of the target at time k given the measurement at time k.

Both the position and the speed estimates make use of the measurement z(k).

The $\alpha - \beta$ filter is not very satisfactory because the coefficients are constants. The chosen values depend upon the problem. In all but the simplest systems such coefficients are functions of time and change as the system is updated. It is difficult to choose values for α and β that are generally suitable and usually the chosen values are not optimal. As we will see, the Kalman filter addresses these issues by changing the gain in line with the measurements.

9 Estimation

This section introduces a number of examples of estimation to illustrate the principles that will lead to the Kalman filter. Some terminology must be introduced:

9.1 Definition of Terms

A number of terms related to estimation are defined below. The terms are illustrated in Figure 20 and Figure 21.

9.1.1 Estimation Algorithm

An estimation algorithm, such as a Kalman filter, uses a series of noisy measurements, taken over past times, to provide an estimate of unknown variables. The estimated values are expected to be more accurate than the raw observations. In some cases, the estimated parameter is not directly observable and its value is inferred from the observations.

9.1.2 Estimate

A filter provides an estimate of an unknown system state based on a series of noisy measurements of related parameters.

9.1.3 Accuracy

Accuracy indicates how close a measurement is to its (possibly unknown) true value. Bias and precision are two components of accuracy.

9.1.4 Bias

A system may be biased in that the measurements have a systematic error. A bias can, for example, add a constant offset to a measurement. Measurement bias is often due to faulty equipment or procedures.

9.1.5 Precision

Precision indicates how the measurement values vary from observation to observation. It is a measure of the scatter of the measurements.

9.1.6 Repeatability and Reproducibility

Accuracy is closely related to bias but precision is a combination of repeatability and reproducibility of the errors in measurements or estimates.

Repeatability refers to variations in measurements when the same procedure is used with the same equipment. Reproducibility refers to when the same process and equipment are used by other people (as in, can they reproduce the results).

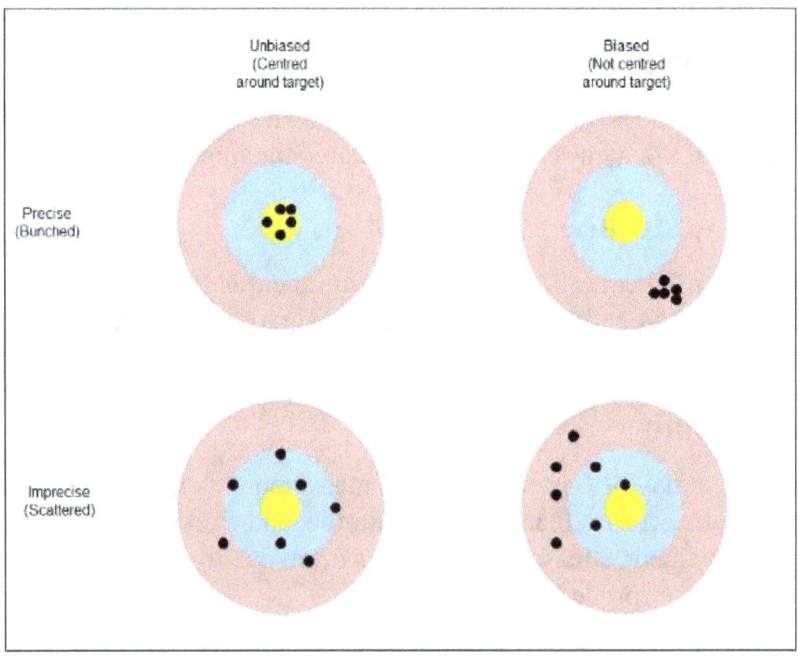

Figure 20 – Estimation Terminology

Kalman Filter

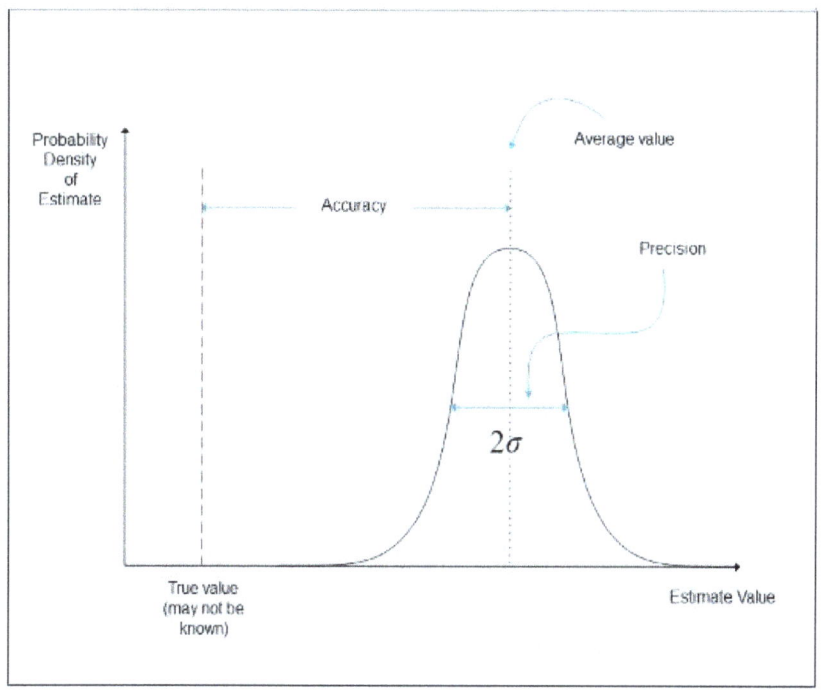

Figure 21 – Estimation Accuracy and Precision

9.2 Recursive Estimation

Consider a constant value corrupted by noise. Samples of the system have values given by

$$y(k) = C + v(k)$$

where v(k) is the uncorrelated measurement noise (random variables) with mean and variance $(0, \sigma_v^2)$.

Measurements may be taken at time intervals k. If N samples are taken then one possible estimate of the value of C is given by the arithmetic mean of stored measurement values:

$$\hat{x}(k) = \frac{1}{N}\sum_{i=0}^{N-1} y(k)$$

Kalman Filter

It is not necessary to store the sample values. If a new measurement is taken then the estimate can be updated as follows:

$$\hat{x}(k+1) = \frac{1}{N+1}\sum_{i=0}^{N} y(k)$$

$$= \frac{1}{N+1}\{\sum_{i=0}^{N-1} y(k) + y(N)\}$$

$$= \frac{1}{N+1}\{N\,\hat{x}(k) + y(N)\}$$

$$= \hat{x}(k) + \frac{1}{N+1}\{y(N) - \hat{x}(k)\}$$

This is a recursive linear estimate of the value of C. It requires no storage of historical data. It requires only the previous estimate and the current measured value y(N). The new estimate consists of the sum of the prior estimate and a weighted contribution from the difference between that prior estimate and the latest measured value.

The current estimate is equal to the previous estimate plus a correction. The correction term corresponds to the deviation of the current estimated value from the latest measurement. This general structure will be seen in the Kalman filter.

9.3 Linear Mean-Squared Estimator

For a signal x we make a linear estimate of the value from m samples y_i as follows:

$$\hat{x} = \sum_{i=1}^{m} h_i y_i$$

We select the m coefficients h to minimise the error

$$E[e^2] = E[(\hat{x} - x)^2]$$

This is expressed as

$$E[e^2] = E[(\hat{x} - x)^2] = E[\sum_{i=1}^{m} h_i y_i - x]^2$$

Minimise this expression with respect to the m parameters h to obtain

$$\frac{\partial E[e^2]}{\partial h_i} = E\left[\frac{\partial e^2}{\partial h_i}\right] = 2E\left[e\frac{\partial e}{\partial h_i}\right] = 2E[ey_i] = 0 \text{ for all } i$$

Kalman Filter

Thus, the optimum choice is given by

$$E[ey_i] = 0$$

which is known as the orthogonality principle.

Substitute for e to give

$$E\left[\left(\sum_{j=1}^{m} h_j j - x\right) y_i\right] = 0$$

$$E\left[\sum_{j=1}^{m} h_j y_j y_i - x y_i\right] = 0$$

$$\left[\sum_{j=1}^{m} h_j R_{ij} - E[x y_i]\right] = 0$$

$$\sum_{j=1}^{m} h_j R_{ij} = g_i$$

where R is the autocorrelation between x_i and x_j and g is the correlation between the random variables x and y.

This can be written as

$$Rh = g$$

and for R positive definite and invertible the solution is

$$h = R^{-1} g$$

This is known as a Weiner filter. Note that this filter requires a knowledge of the number of samples and a calculation of the inverse R matrix. If more data becomes available the computations must be repeated and the R matrix inverted. This is to be contrasted with the previous example, a recursive filter, where data is not stored.

9.4 Fisher Information

Let $p(x, \theta)$ be the joint pdf for random variables x and θ.

Use the following relation

$$\frac{\partial}{\partial \theta} \log(p(x, \theta)) = \frac{1}{p(x,\theta)} \frac{\partial}{\partial \theta} p(x, \theta)$$

to obtain

$$p(x, \theta) \frac{\partial}{\partial \theta} \log(p(x, \theta)) = \frac{\partial}{\partial \theta} p(x, \theta)$$

Kalman Filter

Integrating, we obtain

$$\int_{-\infty}^{+\infty} p(x,\theta) \frac{\partial}{\partial \theta} \log(p(x,\theta)) dx =$$

$$\int_{-\infty}^{+\infty} \frac{\partial}{\partial \theta} p(x,\theta) dx = \frac{\partial}{\partial \theta} \int_{-\infty}^{+\infty} p(x,\theta) dx = \frac{\partial}{\partial \theta}(1) = 0$$

Differentiate again (Leibnitz rule)

$$\int_{-\infty}^{+\infty} \frac{\partial}{\partial \theta} p(x,\theta) \frac{\partial}{\partial \theta} \log(p(x,\theta)) dx +$$

$$\int_{-\infty}^{+\infty} p(x,\theta) \frac{\partial^2}{\partial \theta^2} \log(p(x,\theta)) dx = 0$$

Substitute for $\frac{\partial}{\partial \theta} p(x,\theta)$ to obtain

$$\int_{-\infty}^{+\infty} p(x,\theta) \left[\frac{\partial}{\partial \theta} \log(p(x,\theta))\right]^2 dx +$$
$$\int_{-\infty}^{+\infty} p(x,\theta) \frac{\partial^2}{\partial \theta^2} \log(p(x,\theta)) dx = 0$$

and this is

$$E\left\{\left[\frac{\partial}{\partial \theta} \log(p(x,\theta))\right]^2\right\} = -E\left\{\frac{\partial^2}{\partial \theta^2} \log(p(x,\theta))\right\} \equiv J(\theta)$$

The quantity $J(\theta)$ is called the Fisher information contained in the observable data set x about the unknown parameter θ of the distribution.

The relationship between Fisher information and curvature is illustrated in Figure 22. The curve with high variance has a low curvature and the curve with low variance has high curvature. Intuitively, this seems reasonable because of the presence of the second derivative in the equation for $J(\theta)$ above. Curvature is discussed in the Appendices in section 15.3. The Fisher information is directly related to the curvature of the likelihood function.

Kalman Filter

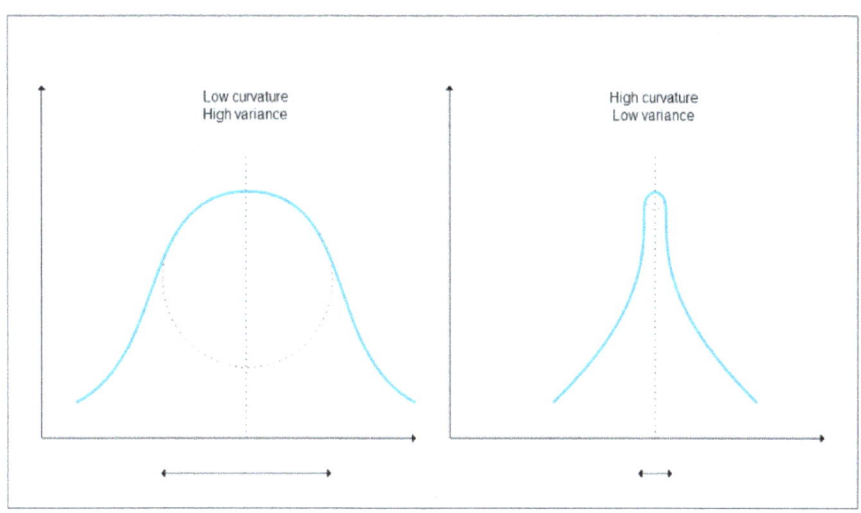

Figure 22 – Curvature and Fisher Information

The term $V = \frac{\partial}{\partial \theta} log(p(x, \theta))$ is known as the score. The expected value of the score is

$$\langle V \rangle = \int_{-\infty}^{+\infty} p(x,\theta) \frac{1}{p(x,\theta)} \frac{\partial}{\partial \theta} p(x,\theta) dx = \int_{-\infty}^{+\infty} \frac{\partial}{\partial \theta} p(x,\theta) dx$$

$$\langle V \rangle = \frac{\partial}{\partial \theta} \int_{-\infty}^{+\infty} p(x,\theta) dx = \frac{\partial}{\partial \theta}(1) = 0$$

Hence, Var(V) = $\langle V^2 \rangle$ and so the Fisher information is the variance of the score. The score is the derivative of the log likelihood function and so setting the score to zero yields maximum likelihood estimate of the parameter.

9.5 Cramer Rao Lower Bound

Let $p(x, \theta)$ be a probability distribution that depends upon a parameter θ. We wish to obtain an estimate of θ from a series of measurements of **x**. What is the minimum variance possible for an unbiased estimator of θ?

An estimator $T(x)$ is said to be an unbiased estimator for θ if

Kalman Filter

$$E[T(x)] = \theta.$$

Using this property, we obtain

$$E[T(x) - \theta] = \int_{-\infty}^{+\infty}[T(x) - \theta]p(x,\theta)dx = 0$$

where $p(x|\theta)$ is the joint pdf for random variables x and θ.

Differentiate wrt θ

$$\int_{-\infty}^{+\infty}[T(x) - \theta]\frac{\partial}{\partial \theta}p(x,\theta)dx - \int_{-\infty}^{+\infty}p(x,\theta)dx = 0$$

The second term is one.

$$\int_{-\infty}^{+\infty}[T(x) - \theta]\frac{\partial}{\partial \theta}p(x,\theta)dx = 1$$

Rewriting by introducing the logarithm term gives

$$\int_{-\infty}^{+\infty}[T(x) - \theta]\sqrt{p(x,\theta)}\sqrt{p(x,\theta)}\frac{\partial}{\partial \theta}\log p(x,\theta)\,dx = 1$$

Now use the Schwarz inequality $(\sum ab)^2 \leq (\sum a^2)(\sum b^2)$ to give

$$1 \leq$$

$$\int_{-\infty}^{+\infty}\left([T - \theta]\sqrt{p(x,\theta)}\right)^2 dx \int_{-\infty}^{+\infty}\left(\sqrt{p(x,\theta)\frac{\partial}{\partial \theta}\log p(x,\theta)}\right)^2 dx$$

Rearranging

$$\int_{-\infty}^{+\infty}(T(x)\theta)^2 p(x,\theta)dx \int_{-\infty}^{+\infty}\left(\frac{\partial}{\partial \theta}\log p(x,\theta)\right)^2 p(x,\theta)dx$$
$$\geq 1$$

Hence, identifying the variance, we obtain

$$Var(T(x)).E\left\{\left(\frac{\partial}{\partial \theta}\log p(x,\theta)\right)^2\right\} \geq 1$$

$$Var(T(x)) \geq \frac{1}{E\left\{\left(\frac{\partial}{\partial \theta}\log p(x,\theta)\right)^2\right\}}$$

From the previous section we see that this involves the Fisher information and this gives

Kalman Filter

$$Var(T(x)) \geq \frac{1}{E\left\{\left(\frac{\partial}{\partial \theta}\log p(x,\theta)\right)^2\right\}} = \frac{-1}{E\left\{\frac{\partial^2}{\partial \theta^2}\log(p(x,\theta))\right\}}$$

The Cramer Rao lower bound is simplified to the inequality

$$Var(T(x)) \geq \frac{1}{J(\theta)}$$

The Cramer Rao inequality states that the mean square error of an unbiased estimator T of θ has a lower bound given by the reciprocal of the Fisher information. Once again, as in 9.4, the curvature is seen to play a part. The relationship between the Cramer Rao lower bound and the curvature of the likelihood function is apparent.

10 Linear Dynamic Systems

The state space approach is based on the concept of a state and is applicable to linear time varying systems. Systems can have multiple inputs and outputs. State space methods utilise vector and matrix notation.

10.1 State

The state of a dynamic system is the smallest set of variables, together with the inputs, that completely determines the behaviour of the system at subsequent times. The state is defined by the set of state variables. The set of 'n' state variables are thought of, and implemented as, the n components of the state vector of dimension n.

10.2 State Space

The state space is the n-dimensional space whose axes consist of the n elements of the state vector (x_1 axis, x_2 axis, etc). Any state can be uniquely represented as a point in the state space.

10.3 Matrix Rank

A matrix of dimension $r \times c$ can be thought of as a set of r row vectors or a set of c column vectors. The rank of the matrix is defined as

1. The maximum number of linearly independent column vectors in the matrix.
2. The maximum number of linearly independent row vectors in the matrix.

The two definitions above are equivalent because, for any matrix, the number of linearly independent columns equals the number of linearly independent rows.

The row rank of a matrix is the number of independent rows; the column rank is the number of independent columns. The row rank and the column rank are always equal.

Kalman Filter

A matrix is said to have full rank if its rank equals the largest possible rank value for a matrix of that size (dimensions). That is, the rank is equal to the number of rows or columns, whichever is smallest.

For a $n \times n$ square matrix, the determinant of the matrix is only non-zero if all the rows and all the columns are linearly independent. Hence, the rank of a $n \times n$ square matrix is n if the determinant is non-zero.

If the matrix is square then full rank implies that:

1. It is invertible.
2. It has a non-zero determinant.
3. It has only non-zero eigen values.

10.4 Observable Systems

A system is observable at t_0 if the output history $y(t), t_0 \leq t \leq t_f \leq \infty$, is sufficient to reconstruct the state $x(t_0)$.

In order to observe the system state, the system must be 'Observable' as defined below.

The above definition is translated into an algebraic condition on the system matrices. A solvable system of linear equations have a solution if and only if the rank of the system matrix is full. For a square matrix, this is equivalent to the requirement that the matrix has a non-zero determinant.

Consider a linear, time invariant, system described by the familiar state space equations for the system and the measurement

$$x(k+1) = \Phi x(k)$$
$$y(k) = Hx(k)$$

where vector **x** has order n, vector **y** has order m, and the matrices are sized accordingly.

The matrixes Φ and H are constant. We must determine if it is possible to identify the value of the state vector given only knowledge of the measurement vector y(k). If we know $x(0) =$

Kalman Filter

x_0 then the system equation provides knowledge of the value of **x** at future times. We thus need to know only x_0 given the measurements **y**(k). The state vector $x(0)$ has n components and so n measurements are required to fix its value.

From the two equations above we can generate n equations for **y**(n) as follows

$$y(0) = Hx(0)$$
$$y(1) = Hx(1) = H\Phi x(0)$$
$$y(2) = Hx(2) = H\Phi x(1) = H\Phi^2 x(0)$$
$$...$$
$$y(n-1) = Hx(n-1) = H\Phi^{n-1} x(0)$$

In matrix notation this is the equation

$$\begin{pmatrix} y(0) \\ y(1) \\ y(2) \\ ... \\ y(n-1) \end{pmatrix} = \begin{pmatrix} H \\ H\Phi \\ H\Phi^2 \\ ... \\ H\Phi^{n-1} \end{pmatrix} x(0)$$

$$nm \times 1 \quad\quad = (nm) \times n \quad\quad n \times 1$$

This linear equation can be solved for **x** by inverting the equation. It has a unique solution for $x(0)$ if and only if the observability matrix, defined below, has rank n. The observability matrix is defined as

$$O = \begin{pmatrix} H \\ H\Phi \\ H\Phi^2 \\ ... \\ H\Phi^{n-1} \end{pmatrix}$$

The rank of the matrix is n if n rows or n columns are linearly independent. For a square matrix this is equivalent to the requirement that the determinant of O is non-zero.

Kalman Filter

The system with measurements defined above is observable if and only if the observability matrix O has rank n (because $n \leq nm$ and so n is full rank).

The algebraic test for observability is not often used in practise. Complex systems may yield complicated algebra and the tests become unwieldy. In practise, it is often the case that systems are analysed and tested thoroughly, with system knowledge, so as to ensure that the states are meaningful and observable. For example, it may be that an unobservable state is initialised and then its value never changes. The value could appear to be sensible if that state variable changes slowly. Such a variable should be identified and eliminated from the state vector.

10.5 Controllable Systems

A system is controllable at time t_o if there is a control history, **u**(t), which in a finite time interval $[t_o, t_f]$, transfers each element of an arbitrary initial state $x(t_o)$ to zero at t_f.

The idea of controllability is similar to that of observability in that it requires a similar algebraic effort, the construction of a controllability matrix, and the evaluation of its row rank. This is specific to control systems and so it is not pursued further. It is mentioned only for completeness.

11 The Kalman Filter

This section presents the derivation of the Kalman filter for the cases of continuous time and discrete time. For the discrete time Kalman filter the derivation is first performed for the easier case of a scalar Kalman filter and then it is repeated for a vector Kalman filter. The Kalman predictor is then derived and discussed.

The Kalman filter is a linear quadratic estimator. It combines the predicted state with noisy measurements to produce optical unbiased estimates of the system states. It is the optimal solution for an estimator for a signal in the presence of independent zero mean white noise sources.

The Kalman filter has been applied in many applications including tracking, navigation, and data prediction. Defining the filter using state space methods facilitates its implementation in discrete systems.

11.1 Random Noise

In signal processing signal noise is usually additive. This means that the observed time series (x) is given by the sum of a signal process (s) and a noise process (n):

$$x(t) = s(t) + n(t)$$

The noise is usually stationary which means that the mean and variance are constant. The autocorrelation and autocovariance depend only upon the difference between two sample times. The autocorrelation is usually assumed to be a delta function – the samples at different times are independent.

At any fixed time, the noise may be a Gaussian random variable.

White noise samples, sampled at different times, are zero mean, have constant finite variance, are uncorrelated with each other, and the spectrum is uniform. If, in addition, the samples at any given time instant are drawn from a Gaussian distribution then the process is Gaussian white noise. Since noise is usually additive

Kalman Filter

then the term Additive White Gaussian Noise (AWGN) is often encountered.

White noise can be defined to satisfy the following constraints:

$$E[w(k)] = 0$$
$$R(n) \equiv E[w(k+n)w(k)] = \sigma^2 \delta(n,0)$$

where R is the autocorrelation function.

Two separate random processes, such as plant noise (w) and measurement noise (v) are usually assumed to be independent of each other. In this case we have the additional relation that their cross-correlation is zero

$$E[w(m)v(n)] = 0 \quad \text{for all m, n}$$

There are other types of noise, such as pink noise. This is characterised by a power dependence that is not constant but instead falls off as the inverse frequency. It is commonly known as "1/f" noise. This noise form is found in electronics and in nature, such as the sound of rainfall.

Another form of noise is an independent and identically distributed (IID) time series. This has all the attributes of white noise with the addition that the elements of the series are independent (not only uncorrelated). Independent means that for two random variables x_1, x_2 the joint pdf is given by

$$p(x_1, x_2) - p(x_1)p(x_2)$$

independence is a stronger condition that whiteness. Independence implies whiteness but whiteness does not imply independence.

11.2 Kalman-Bucy Filter

The continuous time Kalman filter, the Kalman-Bucy filter, assumes a state model and measurement model of the form

$$\frac{dx(t)}{dt} = F(t)x(t) + w(t) \qquad t > 0$$
$$y(t) = H(t)x(t) + v(t)$$

Kalman Filter

$$\tilde{y}(t) = y(t) + v(t)$$

where:

the series $\{w(t)\}$, $\{v(t)\}$ are zero mean white noise processes:

$$E[w(t)] = E[v(t)] = 0$$
(Zero mean)

$$E[w(t)w^T(s)] = Q(t)\delta(t-s)$$
(Finite variance, uncorrelated)

$$E[v(t)v^T(s)] = R(t)\delta(t-s)$$
(Finite variance, uncorrelated)

The series $\{w(t)\}$, $\{v(t)\}$ are uncorrelated:

$$E[w(t)v^T(t)] = 0$$
(Uncorrelated)

x is a vector of the actual states of the system

y is a vector of the actual system output values

ỹ is a vector of the measured process outputs

w is the process noise

v is the measurement noise

The filter aims to provide an estimate $\hat{x}(t)$ of the system state $x(t)$.

The system under observation and the filter are illustrated in Figure 23.

Kalman Filter

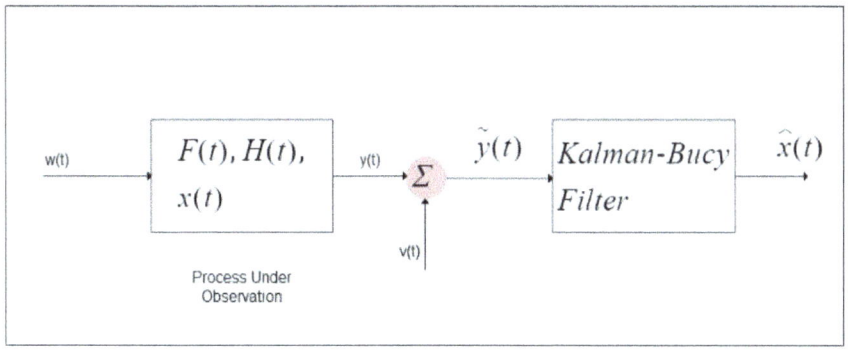

Figure 23 – Kalman-Bucy Filter and Process

The continuous time Kalman filter is illustrated in Figure 24. The filter equations, where time dependence is assumed, are as follows:

$$K = PH^T R^{-1}$$

$$\dot{\hat{x}} = F\hat{x} + K(\tilde{y} - H\hat{x})$$

$$\dot{P} = FP + PF^T - KRK^T + Q$$

where P is the symmetric error covariance estimate and K is the Kalman-Bucy gain. The filter is analogue and requires temporal integration for the continuous updates.

Most Kalman filters are now implemented as digital systems but a continuous time Kalman filter can be implemented as an analogue system. Never the less, this section has been included only for completeness.

Kalman Filter

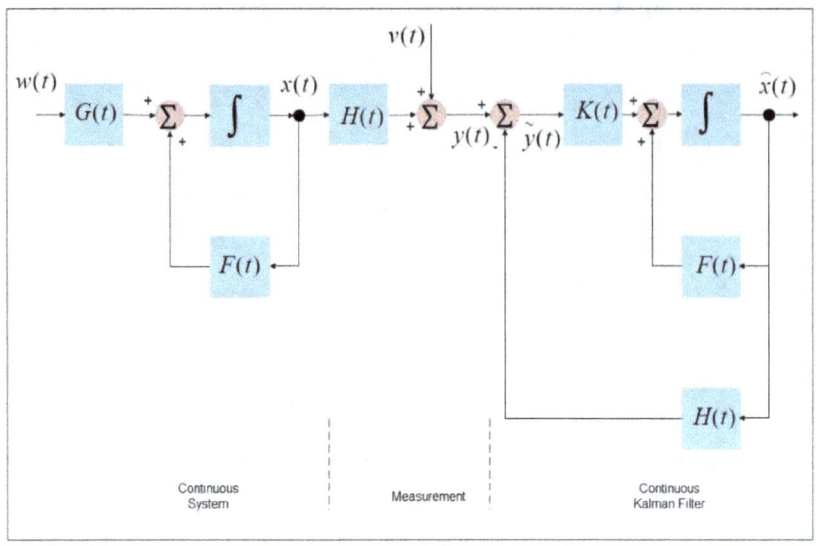

Figure 24 – Continuous Kalman Filter

11.3 Discrete Time Kalman Filter

Consider a scalar system of the general form

$$x(k) = \Phi x(k-1) + u(k-1)$$

with observations

$$y(k) = Hx(k) + v(k)$$

where $u(k)$ and $v(k)$ are zero-mean, finite variance, uncorrelated noise processes. These are known as white noise processes. That is

$$E[u(k)] = E[(v(k)]) = 0$$
$$E[u(i)u(j)] = \sigma_u^2 \delta_{ij}$$
$$E[v(i)v(j)] = \sigma_v^2 \delta_{ij}$$
$$E[u(i)v(j)] = 0$$

In general, the recursive estimator is a weighted sum of the previous estimate and the new observation. The derivation of the discrete Kalman filter is performed in the following sections for

Kalman Filter

the scalar and then the vector cases.

Most Kalman filter systems are implemented on a digital computer (or a microprocessor) and so the most common approach is that of the discrete Kalman filter. Discrete Kalman filters are found in self-driving cars, navigation systems, the GPS system, guidance systems, tracking, etc.

11.3.1   Derivation of Scalar KF

The scalar Kalman filter is derived as an introduction to the more general vector Kalman filter. The calculation is rather long and so it is split into the following manageable blocks.

11.3.1.1   Assumptions

Consider a recursive estimator of the form

$$\hat{x}(k|k) = \alpha(k)\hat{x}(k-1|k-1) + \beta(k)y(k)$$

for the system

$$x(k) = \phi x(k-1) + w(k)$$
$$y(k) = hx(k) + v(k)$$

The two parameters α(k) and β(k) are to be determined. The determination is obtained by minimising the mean squared error. The error is given by (where $x(k)$ is the unknown actual value)

$$e(k) = \hat{x}(k|k) - x(k)$$

and the mean squared error is

$$p(k) = E[e^2(k)]$$

The following expectations are zero because the terms in brackets are products of independent variables.

$$E[e(k)\hat{x}(k-1|k-1)] = 0$$
$$E[e(k)y(k)] = 0$$
$$E[w(k)\hat{x}(k-1|k-1)] = 0$$

Kalman Filter

11.3.1.2 Calculation of Alpha

From the form of the recursive estimator, above, we obtain

$$p(k) = E[(\alpha(k)\hat{x}(k-1|k-1) + \beta(k)y(k) - x(k))^2]$$

Minimising with respect to $\alpha(k)$ and $\beta(k)$ we obtain the conditions

$$\frac{\partial p}{\partial a} =$$
$$2E[(\alpha(k)\hat{x}(k-1|k-1) + \beta(k)y(k) - x(k))\hat{x}(k-1)]$$
$$= 0$$

$$\frac{\partial p}{\partial b} =$$
$$2E[(\alpha(k)\hat{x}(k-1|k-1) + \beta(k)y(k) - x(k))y(k)]$$
$$= 0$$

The first equation becomes

$$E[(\alpha(k)\hat{x}(k-1|k-1))\hat{x}(k-1|k-1)] =$$
$$E\{[x(k) - \beta(k)y(k)]\hat{x}(k-1|k-1)\}$$

Add to LHS the zero value $-a(k)x(k-1) + a(k)x(k-1)$

$$E\{(\alpha(k)\hat{x}(k-1|k-1) - a(k)x(k-1)$$
$$+ a(k)x(k-1))\hat{x}(k-1|k-1)\} =$$
$$E\{[x(k) - \beta(k)y(k)]\hat{x}(k-1|k-1)\}$$

Substitute for $y(k)$

$$a(k)E\{e(k-1)\hat{x}(k-1) + x(k-1)\hat{x}(k-1|k-1)\} =$$
$$E\{[x(k) - \beta(k)hx(k) - \beta(k)v(k)]\hat{x}(k-1|k-1)\}$$

Using the relations

$$E[e(k)\hat{x}(k-1|k-1)] = 0$$
$$E[e(k)y(k)] = 0$$

we obtain

Kalman Filter

$$\alpha(k)E\{x(k-1)\hat{x}(k-1|k-1)\} =$$
$$E\{[x(k) - \beta(k)hx(k)]\,\hat{x}(k-1|k-1)\}$$
$$\alpha(k)E\{x(k-1)\hat{x}(k-1|k-1)\} =$$
$$[1 - \beta(k)h]\,E\{x(k)\hat{x}(k-1|k-1\}]$$

Express x(k) as

$$x(k) = \phi x(k-1) + w(k)$$

and use

$$E[w(k)\,\hat{x}(k-1|k-1)] = 0$$

Factor out $E\{x(k-1)\hat{x}(k-1|k-1)\}$ to give

$$\alpha(k) = \phi[1 - \beta(k)h]$$

The estimator becomes

$$\hat{x}(k|k) =$$
$$\phi\hat{x}(k-1|\,k-1) + \beta(k)[y(k) - \phi h\hat{x}(k-1|k-1)]$$

This equation is the best prediction of $\hat{x}(k)$ plus a correction term proportional to the difference between the latest measurement and the estimate of that observation. The progress so far is illustrated in Figure 25.

Kalman Filter

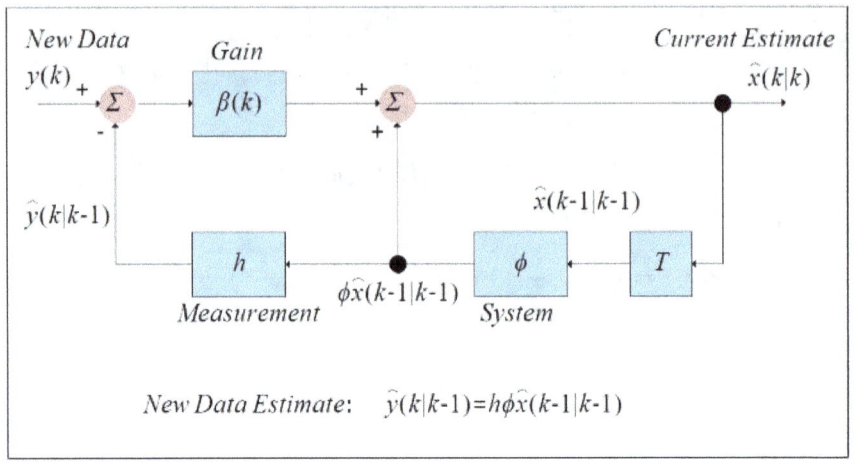

Figure 25 – Scalar Kalman Filter and Beta

The next step is to establish the value of the parameter $\beta(k)$ and then $\alpha(k)$.

11.3.1.3 Calculation of Beta

To derive the form of $\beta(k)$ we proceed from the covariance matrix

$$P(k|k) = E\{e(k|k)e(k|k)\} = E\{e(k)(\hat{x}(k|k) - x(k))\}$$

$$P(k|k) =$$

$$E\{e(k|k)[\alpha(k)\hat{x}(k-1|k-1) + \beta(k)y(k) - x(k)]\}$$

Using the orthogonality relations, we obtain

$$P(k|k) = -E\{e(k|k)x(k)\}$$

From the measurement equation

$$E\{e(k|k)y(k)\} = E\{e(k|k)[hx(k) + v(k)]\}$$

$$0 = hE\{e(k|k)x(k)\} + E\{e(k|k)v(k)\}$$

and then for $P(k|k)$ we obtain

$$P(k|k) = \frac{1}{h}E\{e(k|k)v(k)\}$$

Kalman Filter

Substitute for $e(k|k)$

$$P(k|k) = \frac{1}{h}\{(\hat{x}(k|k) - x(k)v(k)\}$$

$$P(k|k) = \frac{1}{h}\{\alpha(k)\hat{x}(k-1|k-1)v(k) + \beta(k)y(k)v(k) - x(k)v(k)\}$$

Once again, orthogonality relations leave

$$P(k|k) = \frac{1}{h}E\{\beta(k)y(k)v(k)\} =$$

$$\frac{\beta(k)}{h}E\{y(k)v(k)\} = \frac{\beta(k)}{h}E\{v(k)v(k)\}$$

$$\beta(k) = hP(k|k)/\sigma_v^2$$

Now, using the covariance again we have

$$P(k|k) = E\{(\hat{x}(k|k) - x(k))(\hat{x}(k|k) - x(k))\}$$

Using the previous result, substitute for $\hat{x}(k|k)$ to get

$$P(k|k) = E\{(\phi\hat{x}(k-1|k-1) + \beta(k)[y(k) - \phi h\hat{x}(k-1|k-1) - x(k)])^2\}$$

Substitute for y(k) and for x(k), and use orthogonality relations, to obtain,

$$P(k|k) = \{\phi^2(1-\beta(k)h)^2 P(k-1|k-1) + (1-\beta(k)h)^2\sigma_w^2 + \beta(k)^2\sigma_v^2\}$$

Substitute for $P(k|k)$ from above

$$\frac{\beta(k)\sigma_v^2}{h} =$$

$$\{\phi^2(1-\beta(k)h)^2 P(k-1|k-1) + (1-\beta(k)h)^2\sigma_w^2 + \beta(k)^2\sigma_v^2\}$$

Tidying up and collecting terms we obtain finally, the quadratic equation for $\beta(k)$

$$\beta(k)^2[\sigma_v^2 + \phi^2 h^2 P(k-1|k-1) + h^2\sigma_w^2] +$$

$$\beta(k)\left[-2h\phi^2 P(k-1|k-1) - 2h\sigma_w^2 - \frac{\sigma_v^2}{h}\right] +$$

Kalman Filter

$$[\phi^2 P(k-1|k-1) + \sigma_w^2] = 0$$

This equation is discussed in the following section.

11.3.1.4 Solving Quadratic Equation for Beta

The equation for $\beta(k)$ is a quadratic equation of the form

$$ax^2 + bx + c = 0$$

which has standard solutions

$$x = \frac{-b \pm \sqrt{b^2 - 4ac}}{2a}$$

Applying this to the equation developed in the previous section we obtain solutions for $\beta(k)$ as follows:

Let $Q = \phi^2 P(k-1|k-1) + \sigma_w^2$ and then we have for a, b, c:

$$a \rightarrow \sigma_v^2 + h^2 Q$$
$$b \rightarrow -2hQ - \sigma_v^2/h$$
$$c \rightarrow Q$$

And then

$$b^2 - 4ac \rightarrow 4h^2 Q^2 + \frac{\sigma_v^4}{h^2} + 4Q\sigma_v^2 - 4\sigma_v^2 Q - 4h^2 Q^2 = \frac{\sigma_v^4}{h^2}$$

The two solutions are

$$\frac{2hQ + 2\frac{\sigma_v^2}{h}}{2\sigma_v^2 + 2h^2 Q} = \frac{1}{h}$$

$$\frac{2hQ}{2\sigma_v^2 + 2h^2 Q} = h \frac{[\phi^2 P(k-1|k-1) + \sigma_w^2]}{\sigma_v^2 + h^2[\phi^2 P(k-1|k-1) + \sigma_w^2]}$$

The first solution is a constant and is not time varying. Therefore, we require the second solution.

$$\beta(k) = h \frac{[\phi^2 P(k-1|k-1) + \sigma_w^2]}{\sigma_v^2 + h^2[\phi^2 P(k-1|k-1) + \sigma_w^2]}$$

This equation generalises as expected for the vector Kalman filter.

Kalman Filter

11.3.1.5 Solving for Alpha

In section 11.3.1.2 the expression for $\alpha(k)$ in terms of $\beta(k)$ was obtained. The value of $\alpha(k)$ was found to be

$$\alpha(k) = \phi[1 - h\beta(k)]$$

11.3.1.6 The Scalar Kalman Filter

From the above derivation the following summary is useful:

System Model

$$x(k+1) = \phi x(k) + w(k)$$
$$y(k) = hx(k) + v(k)$$

Prediction

$$\hat{x}(k|k-1) = \phi \hat{x}(k-1|k-1)$$
$$\hat{y}(k|k-1) = h\hat{x}(k|k-1)$$

Correction

$$\hat{x}(k|k) = \hat{x}(k|k-1) + \beta(k)[y(k) - h\hat{x}(k|k-1)]$$

Kalman Filter Gain

$$\beta(k) =$$
$$h[\phi^2 p(k-1|k-1) + \sigma_w^2][h^2 \phi^2 p(k-1|k-1) + \sigma_w^2]^{-1}$$

Covariance

$$p(k|k-1) = \phi^2 p(k-1|k-1) + \sigma_w^2$$
$$p(k|k) = [1 - h\beta(k)][\phi^2 p(k-1|k-1) + \sigma_w^2]$$

The application of this system of equations is illustrated in later chapters. The process is illustrated in Figure 26.

Kalman Filter

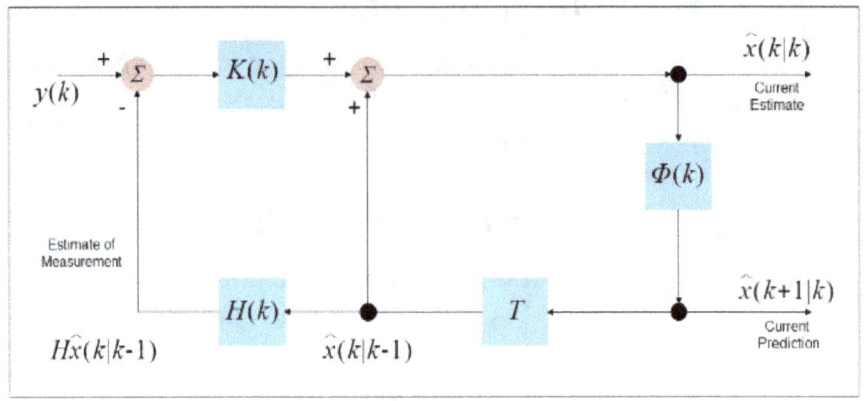

Figure 26 – Scalar Kalman Filter

The scalar Kalman filter equations are next generalised for the case of a higher order system, described by vector equations.

11.3.2 Derivation of Vector KF Equations

The linear system is represented by

$$x(k) = \Phi(k-1) x(k-1) + G(k-1) u(k-1)$$

and we require an estimate $\hat{x}(k|k)$ of the state $x(k)$ at time k, given the previous estimate $\hat{x}(k|k-1)$, and a new measurement $y(k)$ at time k.

The Kalman filter equations are derived by optimising the assumed form of the linear estimator. The recursive estimator is assumed to be of the form

$$\hat{x}(k|k) = \hat{x}(k|k-1) + K(k) [y(k) - H(k)\hat{x}(k|k-1)]$$

where $\hat{x}(k|k-1)$ and $\hat{x}(k|k)$ are the estimates of the state vector $x(k)$ immediately before and after the measurement $y(k)$ at time k. The estimate is corrected at the time of each measurement by the weighted difference of the actual and the forecast measurement vectors. The weight is the Kalman gain $K(k)$, which is to be determined.

From the measurement equation

Kalman Filter

$$y(k) = H(k)x(k) + v(k)$$

and the definitions of the estimation errors $\tilde{x}$

$$\hat{x}(k|k) = x(k) + \tilde{x}(k|k)$$
$$\hat{x}(k|k-1) = x(k) + \tilde{x}(k|k-1)$$

we obtain an expression for the error after the kth measurement

$$\tilde{x}(k|k) = \hat{x}(k|k) - x(k) =$$
$$\hat{x}(k|k-1) + K(k)\,[y(k) - H(k)\hat{x}(k|k-1)] - x(k)$$
$$= \hat{x}(k|k-1) + K(k)\,[y(k) - H(k)\hat{x}(k|k-1)] - \tilde{x}(k|k-1) + \hat{x}(k|k-1)$$
$$= K(k)\,[y(k) - H(k)\hat{x}(k|k-1)] + \hat{x}(k|k-1)$$
$$= K(k)\,[H(k)x(k) + v(k) - H(k)\hat{x}(k|k-1)] + \hat{x}(k|k-1)$$
$$= K(k)\,[H(k)x(k) - H(k)\hat{x}(k|k-1) + v(k)] + \hat{x}(k|k-1)$$
$$= K(k)\,[-H(k)\tilde{x}(k|k-1) + v(k)] + \hat{x}(k|k-1)$$
$$= [I - K(k)H(k)]\,\tilde{x}(k|k-1) + K(k)\,v(k)$$

The error covariance matrix is defined to be

$$P(k|k) - E[\tilde{x}(k|k)\,\tilde{x}(k|k)^T]$$

From the above equation for x̃(k | k) we obtain

$$P(k|k) =$$
$$[I - K(k)H(k)]\,E\{\tilde{x}(k|k-1)\tilde{x}(k|k-1)^T\}\,[I - K(k)H(k)]^T$$
$$+ K(k)\,E\{v(k)v(k)^T\}K(k)^T$$
$$+ (I - K(k)H(k))\,E\{\tilde{x}(k|k-1)v(k)^T\}\,K(k)^T$$
$$+ K(k)\,E\{v(k)\,\tilde{x}(k|k-1)^T\}\,[I - K(k)H(k)]^T$$

The error correlations are given by

$$E\{v(k)\,\tilde{x}(k|k-1)^T\} = 0$$

Kalman Filter

$$E\{\tilde{x}(k|k-1)v(k)^T\} = 0$$

and by definition

$$P(k|k-1) = E\{\tilde{x}(k|k-1)\tilde{x}(k|k-1)^T\}$$

Hence, for the error covariance propagation, we obtain:

$$P(k|k) = [I - K(k)H(k)] P(k|k-1)[I - K(k)H(k)]^T + K(k)R(k)K(k)^T$$

We are now in a position to define an optimum choice for K(k). We can define the optimum value of K(k) to be the value that minimises the trace of the error covariance matrix $P(k|k)$. This is

$$\text{Trace } P(k|k) = E\{\sum_i \tilde{x}_i(k|k)\tilde{x}_i(k|k)\} = E\{\tilde{x}(k|k)^T\tilde{x}(k|k)\}$$

We are minimising the length of the estimation error vector with respect to the gain. Thus, we take the partial derivative of the trace of $P(k|k)$ with respect to K(k) and equate the result to zero.

We must take the partial derivative of the trace of the product of two matrices as follows (where B is symmetric):

$$\frac{\partial}{\partial A}[\text{trace }(A B A^T)] = 2AB$$

Using this relation with the equation above for P(k | k) we obtain

$$-2[I - K(k)H(k)] P(k|k-1) H(k)^T + 2 K(k)R(k) = 0$$

and the result for K(k) is

$$K(k) = P(k|k-1) H(k)^T \{H(k)P(k|k-1)H(k)^T + R(k)\}^{-1}$$

This is the Kalman gain matrix.

Substituting equation for $K(k)$ into the equation for P(k|k-1) gives

$$P(k|k-1) = [I - K(k)H(k)] P(k|k-1)$$

and equation 1 is

Kalman Filter

$$\hat{x}(k|k) = \hat{x}(k|k-1) + K(k)[y(k) - H(k)\hat{x}(k|k-1)]$$

We now have the following system of equations for an update of the state given a new measurement:

Extrapolation:

$$\hat{x}(k|k-1) = \Phi(k-1)\hat{x}(k-1|k-1)$$
$$P(k|k-1) =$$
$$\Phi(k-1)P(k-1|k-1)\Phi(k-1)^T + Q(k-1)$$

Measurement:

Measure **y**(k)

Update:

$$K(k) =$$
$$P(k|k-1)H(k)^T\{H(k)P(k|k-1)H(k)^T + R(k)\}^{-1}$$
$$P(k|k) = [I - K(k)H(k)]P(k|k-1)$$
$$\hat{x}(k|k) = \hat{x}(k|k-1) + K(k)[y(k) - H(k)\hat{x}(k|k-1)]$$

This process is illustrated in Figure 27.

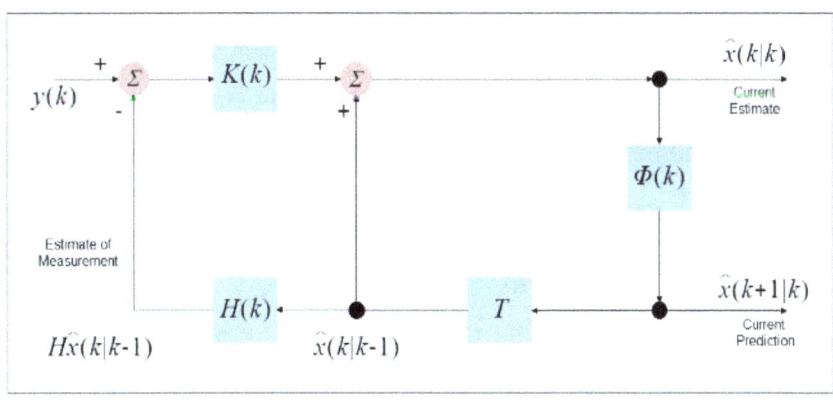

Figure 27 – Kalman Filter

Kalman Filter

11.4 Kalman Predictor

It is often necessary to estimate (predict) the value of a future state x(k) from measurements $y_1, y_2, \ldots, y_m$ where $m < k$. The optimal linear estimator of the state is the Kalman predictor. As we shall see, the Kalman predictor is similar to the Kalman filter.

11.4.1 The Scalar Kalman Predictor Equations

The Kalman predictor equations are derived in a very similar way to the Kalman filter equations. In this section we summarise the results.

We consider the prediction one step into the future. The signal model and measurement model equations are:

$$x(k) = ax(k-1) + w(k-1)$$
$$y(k) = hx(k) + v(k)$$

In contrast to the Kalman filter used to estimate $\hat{x}(k|k)$ we require the estimate (prediction) of the value of x at tine k+1, given data up to the time k. That is denoted by $\hat{x}(k+1|k)$. We require the optimum value, defined to be the value that minimises the mean-square prediction error. Thus, we minimise

$$p(k+1|k) = E[e(k+1|k)e(k+1|k)] =$$
$$E[(x(k+1) - \hat{x}(k+1|k))((x(k+1) - \hat{x}(k+1|k))]$$

We assume the following form for the recursive estimator:

$$\hat{x}(k+1|k) = \alpha(k)\hat{x}(k|k-1) + \beta(k)y(k)$$

This is a weighted sum of the previous prediction and the current measurement. The coefficients α and β are to be determined from the minimisation condition. Minimise $p(k+1|k)$ and proceed as for the Kalman filter.

The Kalman predictor system is summarised below. The system model is unchanged.

Prediction

Kalman Filter

$$\hat{x}(k+1|k) = \phi\hat{x}(k|k-1) + \beta(k)[y(k) - h\hat{x}(k|k-1)]$$

Predictor Gain

$$\beta(k) = \phi h p(k|k-1)[h^2 p(k|k-1) + \sigma_v^2]^{-1}$$

Covariance

$$p(k+1|k) = \phi^2 p(k|k-1) - \phi h \beta(k) p(k|k-1) + \sigma_w^2$$

The Kalman predictor is illustrated in Figure 28.

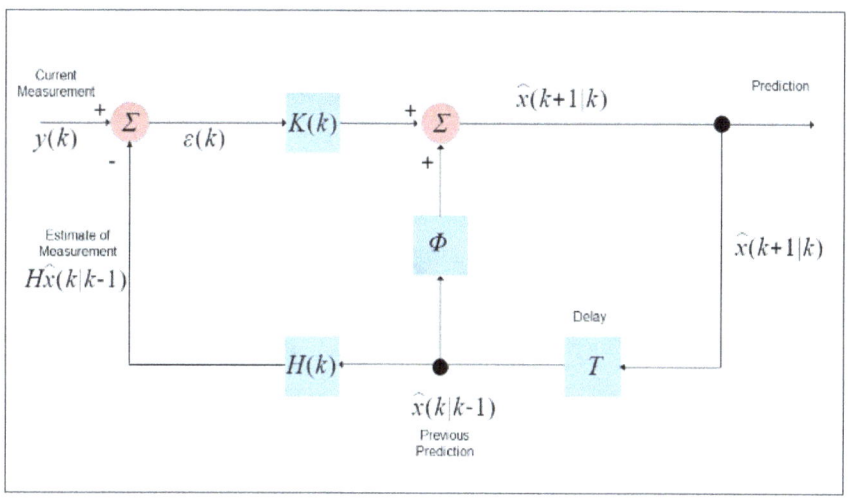

Figure 28 – Scalar Kalman Predictor

11.4.2   The Vector Kalman Predictor

The vector implementation of the Kalman predictor is developed in the same way as the vector Kalman filter.

11.5 Estimation and Prediction

The similarity of the Kalman filter and Kalman predictor equations indicates that both algorithms can be combined in an efficient way. The method can be gleaned from the respective equation sets and the result is illustrated in Figure 29.

Kalman Filter

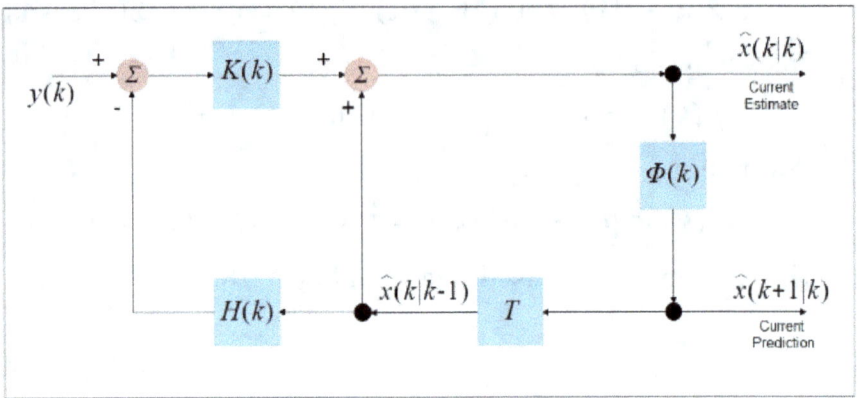

Figure 29 – Simultaneous Estimation and Prediction

12 Implementation

Implementation of the discrete time Kalman filter involves numerical considerations such as rounding errors, efficient techniques for matrix inversion, and algorithms for matrix factorisation. These are discussed below.

12.1 Digital Systems

Digital systems store numbers in a binary representation which could be fixed-point or floating-point. In either case, the representation of the number is finite so that a number such as 1/3 = 0.333... cannot be exactly represented. Modern processors, such as microcontrollers, usually have a floating-point capability and so the treatment of the fixed-point representation is less important than it once was.

Floating-point numbers provide a representation of numbers such a 3.1415..., namely a fractional decimal number, but there is usually a round-off error. Despite the accuracy provided by 32-bit or even 64-bit processors there is still the possibility that round-off error can render a calculation unstable. Small round-off errors can compound and possibly render a matrix non-positive-definite, so that its attempted inversion results in a zero-divide error, and this results in a reset of the processor.

12.2 Roundoff Error

Round-off errors tend to be statistically random and so on average can cancel out. However, there are times when they produce particularly bad effects. For example, if two numbers, close in value, are subtracted, then the result can be catastrophically inaccurate, sometimes to the point that the value zero is returned. If the round-off errors are always of the same sign then the error grows rather than cancels statistically.

It is not possible to predict the effect of round-off errors and so it is necessary to take actions that will eliminate some of their consequences. For example, a matrix can be factored and then the

Kalman Filter

factored components are propagated, and the desired matrix is reassembled after the propagation, in order to guarantee that a matrix remains positive definite and hence invertible. Examples of factored matrices and matrix inversion routines are presented below.

12.3 Positive Definite Matrix

Symmetric positive definite matrices are useful because they can be factored into the product of a symmetric matrix and its transpose and because they can be inverted. Both of these features are important for the implementation of a Kalman filter.

A matrix is positive definite if any of the following tests are valid

1. The matrix is symmetric and all eigenvalues $\lambda > 0$.
2. All upper left determinants of the matrix are > 0.
3. The matrix is symmetric and every pivot is > 0.
4. The matrix P is positive definite iff $P = S^T S$ where S has independent columns.
5. For all non-zero vectors x, $x^T P x > 0$.

Note also that

6. If a matrix is positive definite then it is full-rank.

12.4 Positive Semi-definite Matrix

A matrix is said to be positive semi-definite if all its eigenvalues are non-negative. A positive semi-definite matrix can have eigenvalues that are zero. Such a matrix has zero determinant and therefore cannot be inverted. An example is the zero matrix which has eigenvalues which are all zero. It is positive semi-definite and it cannot be inverted.

12.5 Matrix Inversion

If a matrix P is positive definite then P is invertible. To compute the inverse of a matrix can be computationally expensive. If the

Kalman Filter

matrix is factored into a matrix S (upper triangular) and its transpose (lower triangular) then we have

$$P = S^T S$$

and then

$$P^{-1} = (S^T S)^{-1} = S^{-1}(S^T)^{-1}$$

The triangular matrices can be inverted efficiently and then the inverse of P can be found.

12.6 Errors, Innovations, and Residuals

The innovation is defined as the difference between the current observed state of a variable measurement and the optimal forecast of that value based on previous information. It is given by:

$$i(k) = y(k) - H\hat{x}(k|k-1)$$

The innovation is a measure of the new information provided by a new measurement at time k-1. If the system is performing optimally then the innovation is uncorrelated and appears to be a zero mean, white noise process.

A related term is the residual. In general, for a time-series model, the residual is the difference between the observations and the corresponding fitted values. In the case of the Kalman filter there are two residuals.

The pre-fit residual is identical to the innovation and given by

$$r(k) = y(k) - H\hat{x}(k|k-1)$$

The post-fit residual is given by

$$r(k) = y(k) - H\hat{x}(k|k)$$

Note the following terminology: The error is the deviation of an observed value from the (unobservable) true value, whereas the residual is the difference between the observed value and the estimated value of the observed quantity.

The innovation can be used to monitor the progress of a Kalman filter, as we will see below.

Kalman Filter

12.7 Cholesky Decomposition

In the previous section the decomposition of a matrix into the product of an upper triangular matrix and its transpose was discussed. This can be achieved using Cholesky decomposition as follows.

Every symmetric positive definite matrix P can be factored as

$$P = S^T S$$

where S is upper triangular with positive diagonal elements. S is called the Cholesky factor of P and can be thought of as a 'square root' of the positive definite matrix P. Hence the term square root Kalman filter. The algorithm can be illustrated with an example.

12.7.1 Example: Cholesky Decomposition

The matrix S is given by

$$S = \begin{pmatrix} 1 & -1 & 2 \\ 0 & 2 & -2 \\ 0 & 0 & 3 \end{pmatrix}$$

The matrix P is given by the product

$$P = S^T S = \begin{pmatrix} 1 & -1 & 2 \\ -1 & 5 & -6 \\ 2 & -6 & 17 \end{pmatrix}$$

Note that

> The element P_{11} is 1 and is positive.
>
> The determinant $\begin{vmatrix} 1 & -1 \\ -1 & 5 \end{vmatrix}$ is 4 and is positive.
>
> The determinant of the matrix P is 36 and is positive.

Thus, confirming that P is positive definite as per item 2. in section 12.3 above.

The multiplication can be reversed using the Cholesky decomposition as follows

Kalman Filter

Let P be given by

$$P = \begin{pmatrix} 1 & -1 & 2 \\ -1 & 5 & -6 \\ 2 & -6 & 17 \end{pmatrix} =$$

$$\begin{pmatrix} S_{11} & 0 & 0 \\ S_{12} & S_{22} & 0 \\ S_{13} & S_{12} & S_{33} \end{pmatrix} \begin{pmatrix} S_{11} & S_{12} & S_{13} \\ 0 & S_{22} & S_{23} \\ 0 & 0 & S_{33} \end{pmatrix}$$

The first row of S is given by

$$1 \quad -1 \quad 2 = S_{11}S_{11} \quad S_{11}S_{12} \quad S_{11}S_{13}$$

Hence

$$S_{11} = 1, \ S_{12} = -1, \ S_{13} = 2$$

This leaves the unknown values given by

$$\begin{pmatrix} 1 & -1 & 2 \\ -1 & 5 & -6 \\ 2 & -6 & 17 \end{pmatrix} = \begin{pmatrix} 1 & 0 & 0 \\ -1 & S_{22} & 0 \\ 2 & S_{23} & S_{33} \end{pmatrix} \begin{pmatrix} 1 & -1 & 2 \\ 0 & S_{22} & S_{23} \\ 0 & 0 & S_{33} \end{pmatrix}$$

Multiplying to obtain the element P_{22} gives

$$5 = 1 + S_{22}S_{22} \text{ and thus } S_{22} = 2.$$

Multiplying to obtain the element P_{23} gives

$$-6 = -2 + S_{22}S_{23} = -2 + 2S_{23} \text{ and thus } S_{23} = -2.$$

Finally, multiplying to obtain the element P_{33} gives

$$17 = 4 + S_{23}S_{23} + S_{33}S_{33} = 4 + 4 + S_{33}S_{33} \text{ and thus } S_{33} = 3.$$

The values of the original matrix S have been revealed. This process can be generalised into an algorithm to find the Cholesky factor of any symmetric positive definite matrix.

The product of a matrix and its transpose, AA^T or A^TA, is always symmetric and positive definite and therefore is invertible.

Other forms of matrix decomposition are possible, such as LU decomposition. The Cholesky algorithm, and other factorisation

Kalman Filter

algorithms, are implemented in numerical packages such as the GNU Scientific Library (GSL).

12.8 Covariance Matrix Symmetry

The covariance matrix is symmetric and positive-semi-definite and this property must be maintained during computation. The covariance propagation equations are given by

$$P(k|k-1) = \Phi P(k-1|k-1)\Phi^T + Q(k-1)$$
$$P(k|k) = [1 - K(k)H]P(k|k-1)$$

The first equation guarantees symmetry but the second equation involves a subtraction and can result in loss of symmetry and positive definiteness due to rounding errors. The second equation can be can be developed as follows:

$$K(k) = P(k|k-1)H^T[HP(k|k-1)H^T + R]^{-1}$$
$$P(k|k) = P(k|k-1) - KHP(k|k-1)$$

Multiply the K(k) equation on the right by

$$(HP(k|k-1)H^T + R)K^T$$

to give

$$K(k)(HP(k|k-1)H^T + R)K^T = P(k|k-1)H^TK^T$$

Rearrange

$$KHP(k|k-1)H^TK^T - P(k|k-1)H^TK^T + KRK^T = 0$$

Add to the equation for $P(k|k)$ to give

$$P(k|k) =$$
$$P(k|k-1) - KHP(k|k-1) + KHP(k|k-1)H^TK^T$$
$$-P(k|k-1)H^TK^T + KRK^T$$

$$P(k|k) =$$
$$(1 - KH)P(k|k-1) - (1 - KH)P(k|k-1)H^TK^T + KRK^T$$
$$P(k|k) = (1 - KH)P(k|k-1)(1 - H^TK^T) + KRK^T$$

Kalman Filter

This equation is known as the Joseph form of the error covariance update. The factor with a subtraction is squared, which eliminates the possibility of a negative result. Its quadratic form guarantees that the result is symmetric and positive semi-definite.

Additionally, this version of the covariance matrix update equation not only guarantees symmetry but, provided the matrix $P(k|k-1)$ is positive definite, then the updated matrix $P(k|k)$ will also be positive definite. For this reason, the Joseph form is more stable and robust than the standard update equation and so the Joseph form, given above, is the $P(k|k)$ update equation of choice.

13 Examples

The following examples of Kalman filter implementation explore a number of practical details. In each case the example is described, the results are presented and discussed, and the code is made available.

13.1 A Note on Software

Most implementations of Kalman filters are to be found in real-time, embedded systems in such applications as automotive, aerospace, and control. These systems are usually implemented on a microcontroller, such as the ARM Cortex based series of devices. These devices almost always have an associated 'C' compiler because that is the language most suitable for embedded systems. Note that C++ is not permitted for safety critical systems.

For this reason, the examples discussed below are all implemented in 'C'. The code is supplied for each example. The code is for pedagogical purposes only and has been written to illustrate the detailed linear algebraic manipulations. These are performed using the GNU Scientific Library (GSL).

In a typical application, the Kalman filter manipulations would be hidden behind meaningfully named functions and made available in a static library. For example:

> initialise()
>
> update_system()
>
> update_measurement()
>
> predict_state()
>
> predict_covariance()
>
> compute_kalman_gain()
>
> estimate_state()
>
> estimate_covariance()

Kalman Filter

The library would be linked to the application code. Such an approach has a number of advantages, such as code reuse, data hiding, and encapsulation.

Software engineering is not within our scope and so these details are discussed no further.

13.2 Random Number Generation

In section 4.9 the central limit theorem was discussed with regard to the Gaussian distribution. When multiple sources are added to produce noise then the result is usually a Gaussian distribution and thus, for natural sources of noise, such as process noise (w) and measurement noise (v), it is reasonable to model such noise sources with a Gaussian distribution.

There are many well-known, tried and tested, methods for the generation of a pseudorandom series from a Uniform distribution but, to simulate natural signal noise, it is then necessary to transform the generated series into samples from a Gaussian distribution.

A pseudorandom series is not truly random and will eventually repeat because the word length is finite. However, a computer algorithm (a pseudorandom number generator – PRNG) generated series of numbers may have the appearance of a random number sequence. A pseudo random sequence must have the following properties:

1. The sequence has a large period before repeating.
2. The sequence is reproducible.
3. The sequence satisfies statistical tests of randomness.
4. The sequence is repeatable.
5. The sequence is uniformly distributed on the interval [0, 1].

In practise, a function to generate a series of pseudorandom numbers begins with a user supplied 'seed' and then the sequence is generated recursively from the seed. The seed is usually chosen to be a large prime number. From a given seed the sequence is

Kalman Filter

always the same and so the sequence is reproducible, if required. In some cases, the seed can be generated from the computer 'date and time' to produce a nonrepeatable sequence.

13.2.1   Gaussian Pseudorandom Numbers

If random numbers are produced with a Gaussian distribution and zero mean and unit variance then multiplying each random number by σ gives a series with standard deviation σ. For a non-zero mean then it is necessary only to add the required mean value to each number. Hence, for general use, we require only a series with a Gaussian pdf

$$P(x) = N(0,1) = \frac{1}{\sqrt{2\pi}} exp(-x^2/2)$$

13.2.2   The Box-Muller Method

The Box-Muller transform is used for the generation of pairs of independent, Gaussian distributed, random numbers from a source of uniformly distributed random numbers. We need the inverse transform method to generate the Box-Muller algorithm We state the method but do not prove it. Conceptually, with a change of variables we switch from Cartesian coordinates to polar coordinates, as illustrated in Figure 30. The details are provided in the following section.

Kalman Filter

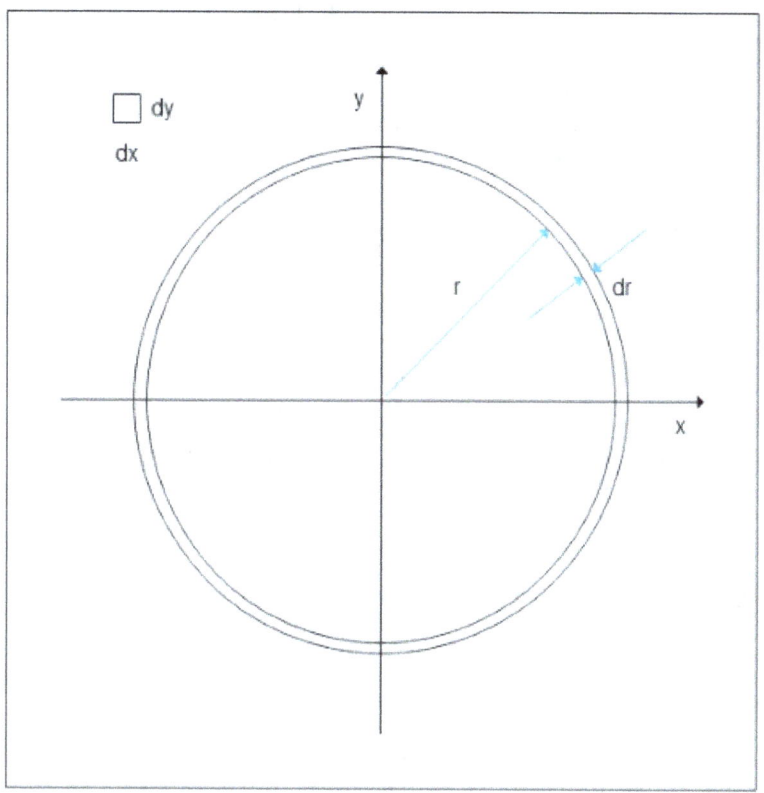

Figure 30 – Box-Muller Method

The Inverse Transform Method

Let $F, x \in R$, denote any cumulative distribution function (CDF). Let $F^{-1}(y), y \in [0, 1]$ denote the inverse function. Define $X = F^{-1}(U)$, where U has the continuous uniform distribution over the interval (0,1). Then X is distributed as F, that is, $P(X \leq x) = F(x), x \in R$.

The steps in the development of the Box-Muller method are summarised as follows

1. Draw two samples from a uniform distribution.
2. The Gaussian distribution does not have an analytic form

Kalman Filter

for the CDF and so the Inverse Transform Method cannot be used directly. Change to polar coordinates to obtain the CDF and then the inverse CDF of a uniform distribution and an exponential distribution.

3. Use the samples from step 1., in the inverse CDF from step 2., to produce samples from a uniform distribution and an exponential distribution.
4. Transform the samples from step 3 to produce samples from a Gaussian distribution.
5. The generation of these samples involves cos and sin functions. To reduce the CPU loads the Box-Muller method is now used. The Box-Muller method is a way to obtain the sin and cos terms from the geometric relations 'sin = opposite/hypotenuse', 'cos = adjacent/hypotenuse'.

The method is now discussed in detail.

Step 1: Generate two uniform random numbers

Generate y_1, y_2 from U(0, 1).

Step 2: Calculate the CDF

To apply the Inverse Transform Method, we require the analytic form of the Gaussian CDF. However, this is given by

$$X(y) = \frac{1}{\sqrt{2\pi}} \int_0^y e^{-t^2/2} \, dt$$

This is known as the error function (erf) and it is not integrable. Instead, to obtain an integrable pdf we transform to polar coordinates. Using the relation

$$p(y_1, y_2) dy_1 dy_2 = p(x_1, x_2) \left| \frac{\partial(y_1, y_2)}{\partial(x_1, x_2)} \right| dx_1 dx_2$$

and the transformation

$$x_1 = y_1 \cos y_2$$
$$x_2 = y_1 \sin y_2$$

and

$$y_1^2 = x_1^2 + x_2^2$$

Kalman Filter

$$y_2 = \tan^{-1}\left(\frac{x_2}{x_1}\right)$$

with the Jacobian equal to

$$\begin{vmatrix} \cos y_2 & -y_1 \sin y_2 \\ \sin y_2 & y_1 \cos y_2 \end{vmatrix} = y_1$$

we obtain

$$p(y_1, y_2) dy_1 dy_2 = \frac{1}{2\pi} dy_2 \, e^{-y_1^2/2} y_1 dy_1$$

Let $u = y_1^2/2$ and then $du = y_1 dy_1$. The integral becomes an exponential pdf:

$$I(x) = \int_0^x e^{-u} du = 1 - e^{-x} = s$$

We obtain the product of a Uniform distribution $U(0, 2\pi)$ and an exponential distribution (with $\lambda = 1$), $E(1)$.

Step 3: Generate the Inverse CDF

This is the exponential distribution (with $\lambda = 1$) CDF and it can be inverted

$$1 - s = e^{-x}$$
$$x = -\ln(1-s)$$

The inverted Uniform distribution CDF $U(0, 2\pi)$ is

$$CDF(\alpha) = \frac{1}{2\pi} \int_0^\alpha d\theta = t$$

$$\alpha = 2\pi t$$

Step 4: Generate samples from a Gaussian distribution

If we supply s in the range [0, 1] then $(1-s)$ is also in the range [1, 0].

$$x_1 = R\cos\theta = \sqrt{-2\ln(s)}\cos 2\pi y_2$$
$$x_2 = R\sin\theta = \sqrt{-2\ln(s)}\sin 2\pi y_2$$
$$s \in (0, 1), y_2 \in [0, 1]$$

The x_1, x_2 pair are the original variables and they are the

Kalman Filter

Gaussian values that we require.

Step 5: Box-Muller Transformation

The trigonometric functions sin and cos are expensive to compute (in CPU time). The Box-Muller transformation is an efficient means of making the computation.

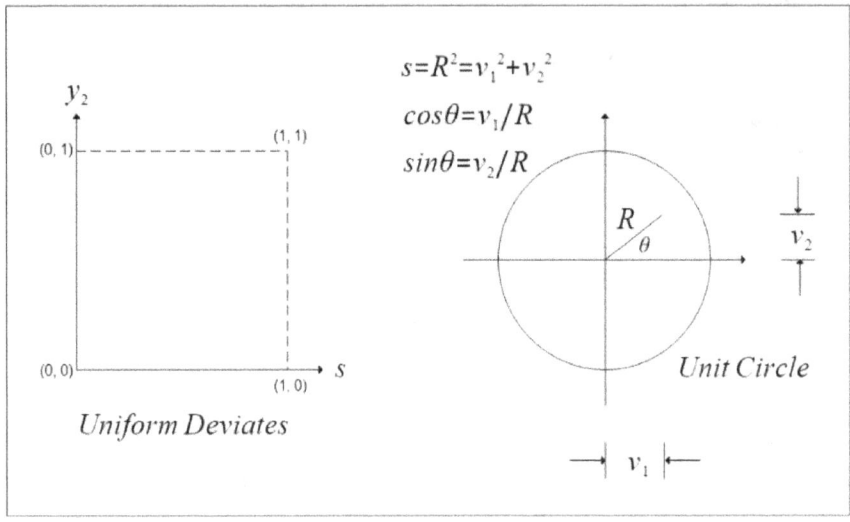

Figure 31 – Box Muller

From Figure 31 we see that we can calculate the trigonometric functions from the geometry of the figure. Provided we chose v_1, v_2 such that $0 < s < 1$, then we can form the following equations for the two Gaussian variables, x and y. The two variables are distributed as N(0, 1), as originally defined. For a given value of R we have

$$x = R\cos\theta = \sqrt{-2\ln s}\, \frac{u_1}{u_1^2+u_2^2} = u_1\sqrt{\frac{-2\ln s}{s}}$$

$$y = R\sin\theta = \sqrt{-2\ln s}\, \frac{u_2}{u_1^2+u_2^2} = u_2\sqrt{\frac{-2\ln s}{s}}$$

Hence for uniformly distributed u values we can generate the

Kalman Filter

required Gaussian distribution of x, y variables.

The algorithm

Now take two uniformly distributed random numbers from U(0, 1). Let the two values be u_1, u_2 and select them such that $s = u_1^2 + u_2^2 \leq 1$. The value of s is also drawn from U(0, 1). From the previous equation, R is found such that the probability of the area under the curve $P(r)$ from 0 to R is s. Then

$$R = \sqrt{-2\ln(1-s)}$$

But, both u and $(1 - u)$ are uniformly distributed on $[0, 1]$. Hence, we can take $R = \sqrt{-2\ln(s)}$.

This algorithm is implemented in the function `r4_normal_01(int *seed)` which is quoted in section 13.8.1 of the Appendices.

The algorithm is summarised as follows:

Gaussian Random Number Generation

Input:

Two uniformly distributed random numbers

$$u_1, u_2 \in (0, 1]$$

Output:

Two independent Gaussian distributed numbers

$$x_1, x_2 \in N(0, 1)$$

Set $a = \sqrt{-2\ln u_1}$

Set $b = 2\pi u_2$

$x_1 = acos(b)$

$x_2 = asin(b)$

return (x_1, x_2)

Kalman Filter

13.3 Position Estimate

This example is selected because it is a simple example of a Kalman filter. It is a scalar Kalman filter, and estimates only the value of a constant in the presence of noise.

13.3.1 Description

The system is described by the following equations:

$$x = C + w$$

The measurements are given by

$$y = x + v$$

where C is a constant value and w and v are both Gaussian random variables which represent signal and measurement noise respectively.

13.3.2 Results

The results of the simulation are displayed in Figure 32. It can be seen that the values of the system variable and its estimate are in good agreement. The error between the estimate and the (unknown in a physical system) real value is small and zero mean. The covariance is initially large and then quickly drops to a value consistent with the error size. These results are typical of a Kalman filter estimator.

Kalman Filter

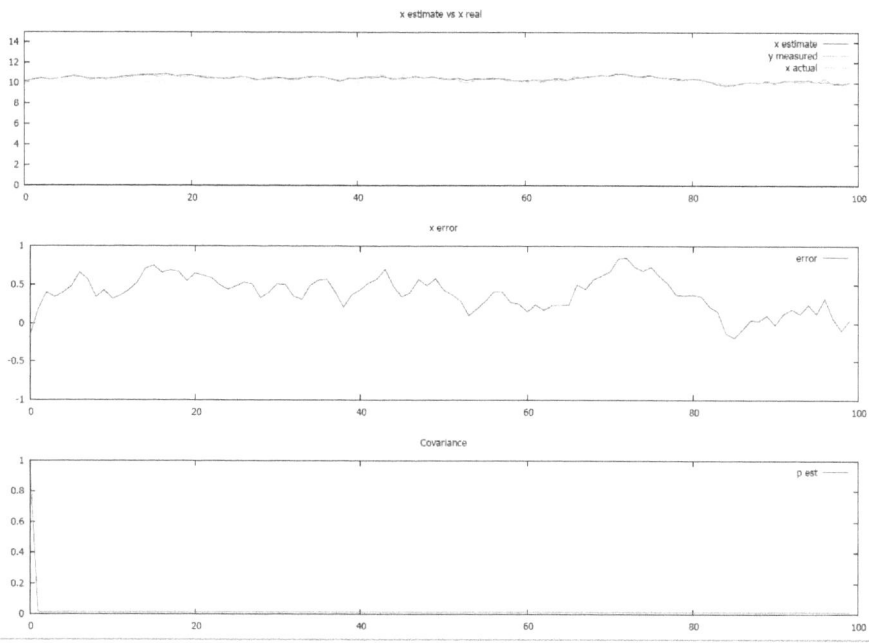

Figure 32 – Scalar KF Results

13.3.3 Code

The code presented below corresponds to the above example. It consists only of initialisation, a main loop, and two file handling functions.

```
/*
===================================================
  Name         : ScalarExample.c
  Author       : Kalman Filter: Introduction
  Version      : 1.0
  Copyright    : Free to use
  Description  : Scalar Kalman filter.
===================================================
*/
```

Kalman Filter

```c
/***************************************************
      Includes
***************************************************/

#include <stdio.h>
#include <stdlib.h>
#include <string.h>
#include <math.h>

#include <complex.h>
#include <normal.h>
#include <time.h>

/***************************************************
      Constants
***************************************************/

#define STDEV_V   0.1
#define STDEV_W   0.1

#define VAR_V    (STDEV_V * STDEV_V)
#define VAR_W    (STDEV_W * STDEV_W)

#define K_MAX    100

/***************************************************
      Typedefs
***************************************************/

/***************************************************
      Types
***************************************************/

/***************************************************
      Globals
***************************************************/

/***************************************************
      Locals
***************************************************/
```

Kalman Filter

```c
int seed = 0;

double phi = 1.0;
double D = 0.0;
double H = 1.0;
double p_pred = 100.0 * VAR_W;
double p_est = 100.0 * VAR_W;
double x_pred = 0.0;
double x_est = 0.0;
double Q = VAR_W;
double R = VAR_V;
double K = 0.0;

FILE* F_OUT;

char file_out[100] =
"C:\\Users\\andre\\Documents\\Books";

/************************************************
     Functions
************************************************/

void open_file(void)
{
  strcat(file_out, "\\Plot\\Scalar_KF.dat");

  F_OUT = fopen(file_out, "w");
}

void close_file(void)
{
     fclose(F_OUT);
}

/************************************************
     Main
************************************************/

int main(void)
{
  puts("New Bias");
```

Kalman Filter

```c
open_file();

fprintf(F_OUT, "Scalar KF\n");

seed = time(0);

/* Initialise */
double x_real = 10.0;
double x_old = 10.0;
double x = 10.0;
double y = 0.0;

x_est = 0.0;

float t = 0.0;

float innov[K_MAX] = {0.0};

fprintf(F_OUT,"p_est %f\n", p_est);
fprintf(F_OUT,"k, x, y, x_est,
        innov, p_est, e\n");

for (int k = 0; k < K_MAX; k++)
{
    t = t + 0.1;

    // Noise
    float w = r4_normal_01(&seed) * STDEV_W;
    float v = r4_normal_01(&seed) * STDEV_V;

    /* Advance system */
    x = phi * x_old + w;

    /* Measurement */
    y = x + v;

    /* Prediction */
    x_pred = phi * x_est;

    p_pred = phi * p_est * phi + Q;
```

Kalman Filter

```c
        /* Innovation process */
        innov[k] = y - H * x_pred;

        /* Kalman gain */
        D =  (H * p_pred * H + R);
        K = p_pred * H / D;

        /* Update (correction) */
        x_est = x_pred + K * (y - H * x_pred);

        p_est = p_pred - K * H * p_pred;

        x_old = x;

        /* Error */
        float e = x_est - x_real;

        /* Output results */
        fprintf(F_OUT, "%d", k);
        fprintf(F_OUT, " %6.2f", x);
        fprintf(F_OUT, " %6.2f", y);

        fprintf(F_OUT, " %6.2f", x_est);
        fprintf(F_OUT, " %6.2f", innov[k]);
        fprintf(F_OUT, " %6.2f", p_est);

        fprintf(F_OUT, " %6.2f\n", e);
    }

    close_file();

    puts("End");

    return EXIT_SUCCESS;
}
```

13.4 Biased Measurement Failure

It is possible to estimate a measurement bias in a system. The bias can then be compensated to maintain an accurate estimate of the

Kalman Filter

system state. In this example the system equations are configured to estimate a bias and then it is shown that the system is not observable and hence that the measurement bias cannot be compensated.

13.4.1 Description

The system is given by

$$x(k+1) = x(k) + w(k)$$

and measurement, with a constant bias b, by

$$y(k) = x(k) + b + v(k)$$

If we augment these equations to estimate the bias (as x2) we obtain

$$\begin{pmatrix} x1(k+1) \\ x2(k+1) \end{pmatrix} = \begin{pmatrix} 1 & 0 \\ 0 & 1 \end{pmatrix} \begin{pmatrix} x1(k) \\ x2(k) \end{pmatrix} + \begin{pmatrix} w1(k) \\ w2(k) \end{pmatrix}$$

$$y(k) = \begin{pmatrix} 1 & 1 \end{pmatrix} \begin{pmatrix} x1(k) \\ b \end{pmatrix} + v1(k)$$

The observation matrix can be calculated to give

$$O = \begin{pmatrix} H \\ H\Phi \end{pmatrix}$$

which is, from the system and measurement equations above,

$$O = \begin{pmatrix} 1 & 1 \\ 1 & 1 \end{pmatrix}$$

The rank of O is not 2 and equivalently

$$\det(O) = 0$$

Hence, the system is not observable and we cannot estimate the bias.

13.5 Biased Measurement

This example illustrates the estimate of an unknown bias in a measurement. The bias can therefore be absorbed into the model and taken into account by the estimator, so that the estimated

Kalman Filter

value is accurate despite the measurement bias. Even if the bias is changing the estimated value can be compensated accordingly.

13.5.1 Description

We now consider a more general system and determine the conditions for observability. The system is given by

$$x(k+1) = \Phi x(k) + w(k)$$
$$y(k) = x(k) + b + v(k)$$

where:

$$\Phi = \begin{pmatrix} \alpha & \beta \\ \gamma & \delta \end{pmatrix}$$

If we augment these equations, as above, we obtain

$$\begin{pmatrix} x1(k+1) \\ x2(k+1) \\ x3(k+1) \end{pmatrix} = \begin{pmatrix} \alpha & \beta & 0 \\ \gamma & \delta & 0 \\ 0 & 0 & 1 \end{pmatrix} \begin{pmatrix} x1(k) \\ x2(k) \\ x3(k) \end{pmatrix} + \begin{pmatrix} w1(k) \\ w2(k) \\ w3(k) \end{pmatrix}$$

$$y(k) = \begin{pmatrix} 1 & 0 & 1 \end{pmatrix} \begin{pmatrix} x1(k) \\ x2(k) \\ x3(k) \end{pmatrix} + v(k)$$

The observability matrix is

$$O = \begin{pmatrix} 1 & 0 & 1 \\ \alpha & \beta & 1 \\ \alpha^2 + \beta\gamma & \alpha\beta + \beta\delta & 1 \end{pmatrix}$$

The first columns and the last column must be linearly independent, hence

$$\alpha \neq 1$$

or

$$\alpha^2 + \beta\gamma \neq 1$$

In terms of the determinant

$$\text{Det}(O) = (1)(\beta - \alpha\beta - \beta\delta) +$$

Kalman Filter

$$(1)(\alpha^2\beta + \alpha\beta\delta - \beta\alpha^2 - \beta^2\gamma) \neq 0$$
$$\beta - \alpha\beta - \beta\delta + \alpha^2\beta + \alpha\beta\delta - \beta\alpha^2 - \beta^2\gamma \neq 0$$

If α is one then

$$\beta - \beta - \beta\delta + \beta + \beta\delta - \beta - \beta^2\gamma = -\beta^2\gamma \neq 0$$

In our example we take

$$\Phi = \begin{pmatrix} \alpha & \beta & 0 \\ \gamma & \delta & 0 \\ 0 & 0 & 1 \end{pmatrix}$$

with

$$\alpha = 1/2, \beta = 1, \gamma = 1/4, \delta = 0$$

These values satisfy the observability constraints. The results are given in the following section.

13.5.2 Results

The results from the simulation runs are presented in Figure 33. It can be seen from the second figure that the bias estimate is accurate and converges to a value close to the real value of 0.5.

Kalman Filter

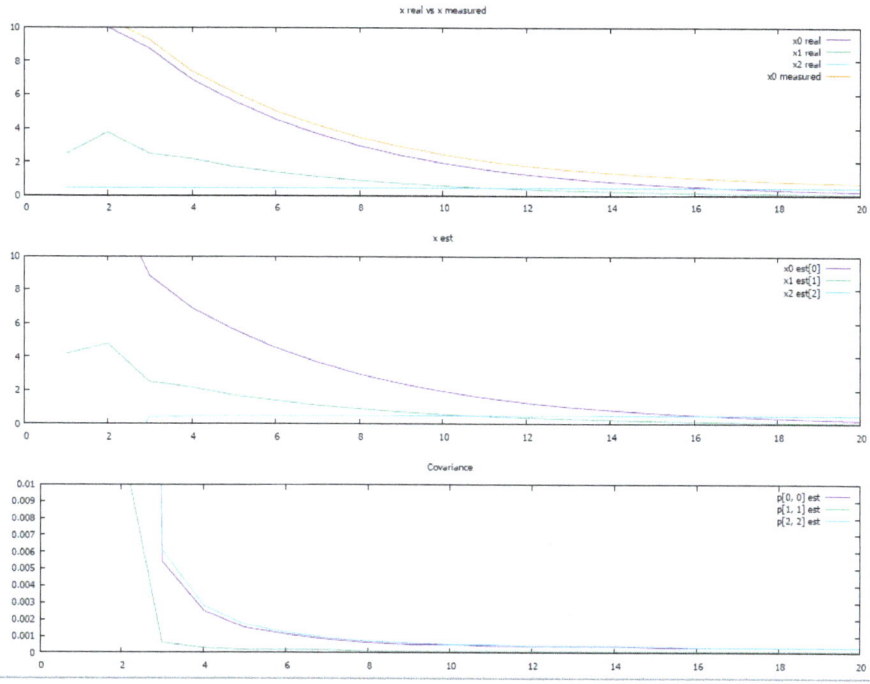

Figure 33 – Bias Estimate Results

13.5.3 Code

The code presented below corresponds to the above example. It consists only of initialisation, a main loop, and two file handling functions.

```
/*
=================================================
  Name        : BiasEstimation.c
  Author      : Kalman Filter: Introduction
  Version     : 1.0
  Copyright   : Free to use
  Description : Bias estimation Kalman filter.
=================================================
*/
```

Kalman Filter

```
/***************************************************
      Includes
***************************************************/

#include <stdio.h>
#include <stdlib.h>
#include <math.h>
#include <string.h>

#include <complex.h>
#include <normal.h>
#include <time.h>

#include <gsl/gsl_matrix.h>
#include <gsl/gsl_blas.h>
#include <gsl/gsl_matrix_double.h>
#include <gsl/gsl_linalg.h>

#include <linear_algebra.h>

/***************************************************
    Constants
***************************************************/

#define STDEV_U   0.01
#define STDEV_V   0.01
#define STDEV_W   0.01

#define VAR_U    (STDEV_U * STDEV_U)
#define VAR_V    (STDEV_V * STDEV_V)
#define VAR_W    (STDEV_W * STDEV_W)

#define X_SIZE 3
#define Y_SIZE 1

#define K_MAX   50

/***************************************************
    Typedefs
***************************************************/
```

Kalman Filter

```
/*************************************************
    Types
*************************************************/

/*************************************************
    Globals
*************************************************/

/*************************************************
    Locals
*************************************************/

static int seed = 0;

static double alpha = 0.5;
static double beta = 1.0;
static double gamma = 0.25;
static double delta = 0.0;

static double T = 0.1;

FILE* F_OUT;

char file_out[100] =
"C:\\Users\\andre\\Documents\\Books";

/*************************************************
    Functions
*************************************************/

void open_file(void)
{
  strcat(file_out,
"\\Plot\\BiasEstimation_KF.dat");

  F_OUT = fopen(file_out, "w");
}

void close_file(void)
{
  fclose(F_OUT);
```

Kalman Filter

```
}

/***********************************************
        Main
***********************************************/

int main(void)
{
  puts("Bias Estimation");

  open_file();

  fprintf(F_OUT, "k, x real[0],
  x_real[1],x_real[2], y, x_est[0], x_est[1],
  x_est[2], p_est p_est p_est\n");

  seed = time(0);

  gsl_matrix* PHI =
      gsl_matrix_alloc(X_SIZE, X_SIZE);
  gsl_matrix_set(PHI, 0, 0, alpha);
  gsl_matrix_set(PHI, 0, 1, beta);
  gsl_matrix_set(PHI, 0, 2, 0.0);
  gsl_matrix_set(PHI, 1, 0, gamma);
  gsl_matrix_set(PHI, 1, 1, delta);
  gsl_matrix_set(PHI, 1, 2, 0.0);
  gsl_matrix_set(PHI, 2, 0, 0.0);
  gsl_matrix_set(PHI, 2, 1, 0.0);
  gsl_matrix_set(PHI, 2, 2, 1.0);

  gsl_matrix* H    =
      gsl_matrix_alloc(Y_SIZE, X_SIZE);
  gsl_matrix_set(H, 0, 0, 1.0);
  gsl_matrix_set(H, 0, 1, 0.0);
  gsl_matrix_set(H, 0, 2, 1.0);

  gsl_matrix* Pest  =
      gsl_matrix_alloc(X_SIZE, X_SIZE);
  gsl_matrix_set(Pest, 0, 0, 1.0);
  gsl_matrix_set(Pest, 0, 1, 0.0);
  gsl_matrix_set(Pest, 0, 2, 0.0);
```

Kalman Filter

```
gsl_matrix_set(Pest, 1, 0, 0.0);
gsl_matrix_set(Pest, 1, 1, 1.0);
gsl_matrix_set(Pest, 1, 2, 0.0);
gsl_matrix_set(Pest, 2, 0, 0.0);
gsl_matrix_set(Pest, 2, 1, 0.0);
gsl_matrix_set(Pest, 2, 2, 1.0);

gsl_matrix* Ppred   =
   gsl_matrix_alloc(X_SIZE, X_SIZE);
gsl_matrix_set(Ppred, 0, 0, 1.0);
gsl_matrix_set(Ppred, 0, 1, 0.0);
gsl_matrix_set(Ppred, 0, 2, 0.0);
gsl_matrix_set(Ppred, 1, 0, 0.0);
gsl_matrix_set(Ppred, 1, 1, 1.0);
gsl_matrix_set(Ppred, 1, 2, 0.0);
gsl_matrix_set(Ppred, 2, 0, 0.0);
gsl_matrix_set(Ppred, 2, 1, 0.0);
gsl_matrix_set(Ppred, 2, 2, 1.0);

gsl_matrix* Q   =
   gsl_matrix_alloc(X_SIZE, X_SIZE);
gsl_matrix_set(Q, 0, 0, VAR_W);
gsl_matrix_set(Q, 0, 1, 0.0);
gsl_matrix_set(Q, 0, 2, 0.0);
gsl_matrix_set(Q, 1, 0, 0.0);
gsl_matrix_set(Q, 1, 1, VAR_W);
gsl_matrix_set(Q, 1, 2, 0.0);
gsl_matrix_set(Q, 2, 0, 0.0);
gsl_matrix_set(Q, 2, 1, 0.0);
gsl_matrix_set(Q, 2, 2, VAR_W);

gsl_matrix* K   =
   gsl_matrix_alloc(X_SIZE, Y_SIZE);
gsl_matrix_set_all(K, 0.0);

gsl_matrix* R   =
   gsl_matrix_alloc(Y_SIZE, Y_SIZE);
gsl_matrix_set_all(R, VAR_V);

gsl_matrix* D   =
   gsl_matrix_alloc(Y_SIZE, Y_SIZE);
```

Kalman Filter

```
    gsl_matrix_set_all(D, 0.0);

    gsl_matrix* Dinv  =
        gsl_matrix_alloc(Y_SIZE, Y_SIZE);
    gsl_matrix_set_all(Dinv, 0.0);

    gsl_vector* Xest = gsl_vector_alloc(X_SIZE);
    gsl_vector* Xpred = gsl_vector_alloc(X_SIZE);
    gsl_vector* X = gsl_vector_alloc(X_SIZE);
    gsl_vector* Xold = gsl_vector_alloc(X_SIZE);

    /* Initialise */
    gsl_vector_set(Xold, 0, 10.0);
    gsl_vector_set(Xold, 1, 10.0);
    gsl_vector_set(Xold, 2, 0.0);

    gsl_vector_set(Xest, 0, 20.0);
    gsl_vector_set(Xest, 1, 20.0);
    gsl_vector_set(Xest, 2, 0.0);

    double x1 = 10.0;
    double x2 = 10.0;
    double x3 = 0.0;

    double bias = 0.5;

    double t = 0.0;

    for (int k = 1; k < K_MAX; k++)
    {
        t = t + T;

        /* Noise */
        float u = r4_normal_01(&seed) * STDEV_U;
        float w = r4_normal_01(&seed) * STDEV_W;
        float v = r4_normal_01(&seed) * STDEV_V;

        // Actual system -----------------------

        /* System */
        double x1_temp = x1;
```

Kalman Filter

```c
double x2_temp = x2;
x1 = alpha * x1_temp + beta * x2_temp;
x2 = gamma * x1_temp + delta * x2_temp;
x3 = bias;

/* x = phi * x_old + w; */
gsl_blas_dgemv(
CblasNoTrans, 1.0, PHI, Xold, 0.0, X);
gsl_vector* w_temp =
    gsl_vector_alloc(X_SIZE);
gsl_vector_set(w_temp, 0, w);
gsl_vector_set(w_temp, 1, u);
gsl_vector_set(w_temp, 2, v);
gsl_vector_add(X, w_temp);

/* Measurement */
/* y = x + bias + v */
gsl_vector* v_temp =
    gsl_vector_alloc(Y_SIZE);
gsl_vector_set(v_temp, 0, v);
gsl_vector* y =
    gsl_vector_alloc(Y_SIZE);
gsl_vector* y_bias =
    gsl_vector_alloc(Y_SIZE);
gsl_vector_set(y_bias, 0, bias);
gsl_vector_set(y, 0, x1);
gsl_vector_add(y, y_bias);
gsl_vector_add(y, v_temp);

// Prediction---------------------------

/* x_pred = phi * x_est; */
gsl_blas_dgemv(
CblasNoTrans, 1.0, PHI, Xest,
0.0, Xpred);

/* p_pred = phi * p_est * phi' + Q; */
gsl_matrix* Temp   =
    gsl_matrix_alloc(X_SIZE, X_SIZE);
gsl_blas_dgemm(
CblasNoTrans, CblasNoTrans,
```

Kalman Filter

```
                 1.0, PHI, Pest, 0.0, Temp);
                 gsl_blas_dgemm(
                 CblasNoTrans, CblasTrans,
                 1.0, Temp, PHI, 0.0, Ppred);
                 gsl_matrix_add(Ppred, Q);

                 // Kalman Gain ------------------------

                 /* D =  (H * p_pred * H' + R); */
                 gsl_matrix* Dtemp  =
                    gsl_matrix_alloc(Y_SIZE, X_SIZE);
                 gsl_matrix* Rtemp  =
                    gsl_matrix_alloc(Y_SIZE, Y_SIZE);
                 gsl_blas_dgemm(
                 CblasNoTrans, CblasNoTrans,
                 1.0, H, Ppred, 0.0, Dtemp);
                 gsl_blas_dgemm(
                 CblasNoTrans, CblasTrans,
                 1.0, Dtemp, H, 0.0, Rtemp);
                 gsl_matrix_add(Rtemp, R);
                 Dinv = matrix_inverse(Rtemp, Y_SIZE);

                 /* K = p_pred * H' / D; */
                 gsl_matrix* Ktemp =
                 gsl_matrix_alloc(X_SIZE, Y_SIZE);
                 gsl_blas_dgemm(
                  CblasNoTrans, CblasTrans,
                  1.0, Ppred, H, 0.0, Ktemp);
                 gsl_blas_dgemm(CblasNoTrans,
                 CblasNoTrans, 1.0, Ktemp, Dinv, 0.0, K);

                 // Update ------------------------------

                 /* x_est =
                    x_pred + K * (y - H * x_pred); */
                 gsl_vector* Ypred  =
                 gsl_vector_alloc(Y_SIZE);
                 gsl_vector* Xtemp  =
                 gsl_vector_alloc(X_SIZE);
                 gsl_vector* Innov  =
                  gsl_vector_alloc(Y_SIZE);
```

Kalman Filter

```c
gsl_vector_memcpy(Innov, y);
gsl_blas_dgemv(
 CblasNoTrans,
 1.0, H, Xpred, 0.0, Ypred);
gsl_vector_sub(Innov, Ypred);
gsl_blas_dgemv(
 CblasNoTrans,
 1.0, K, Innov, 0.0, Xtemp);
gsl_vector_add(Xpred, Xtemp);
gsl_vector_memcpy(Xest, Xpred);

/* p_est = p_pred - K * H * p_pred; */
gsl_matrix* Ktemp1 =
gsl_matrix_alloc(X_SIZE, X_SIZE);
gsl_matrix* Ptemp1 =
gsl_matrix_alloc(X_SIZE, X_SIZE);
gsl_blas_dgemm(
 CblasNoTrans, CblasNoTrans,
 1.0, K, H, 0.0, Ktemp1);
gsl_blas_dgemm(
CblasNoTrans, CblasNoTrans,
1.0, Ktemp1, Ppred, 0.0, Ptemp1);
gsl_matrix_sub(Ppred, Ptemp1);
gsl_matrix_memcpy(Pest, Ppred);

/* x_old = x; */
gsl_vector_memcpy(Xold, X);

// Results ----------------------------

/* Output results */
fprintf(F_OUT, "%d", k);
fprintf(F_OUT, " %6.2f", x1);
fprintf(F_OUT, " %6.2f", x2);
fprintf(F_OUT, " %6.2f", x3);

fprintf(F_OUT, " %6.2f",
    gsl_vector_get(y, 0));

fprintf(F_OUT, " %6.2f",
    gsl_vector_get(Xest, 0));
```

Kalman Filter

```
            fprintf(F_OUT, " %6.2f",
                gsl_vector_get(Xest, 1));
            fprintf(F_OUT, " %6.2f",
                gsl_vector_get(Xest, 2));

            fprintf(F_OUT, " %6.4f",
                gsl_matrix_get(Pest, 0, 0));
            fprintf(F_OUT, " %6.4f",
                gsl_matrix_get(Pest, 1, 1));
            fprintf(F_OUT, " %6.4f\n",
                gsl_matrix_get(Pest, 2, 2));
    }

    close_file();

    puts("End");

    return EXIT_SUCCESS;
}
```

13.6 Falling Body – with Driving Force

This example illustrates a system undergoing the effect of an external force. In this case, gravity.

13.6.1   Description

The system is described by the process equation:

$$x(k + 1) = \Phi x(k) + \Gamma u(k) + w(k)$$

and the measurement equation

$$y(k) = Hx(k) + v(k)$$

where the **x** and **y** vectors are the state vector and measurement vector respectively. The vector **u** is a force term (gravity). The noise terms **w** and **v** are Gaussian and uncorrelated.

The system is illustrated in Figure 34. The matrices Φ, H and Γ have the values

Kalman Filter

$$\Phi = \begin{pmatrix} 1.0 & T \\ 0.0 & 1.0 \end{pmatrix}, \; H = (1.0 \; \; 0.0), \text{ and } \Gamma = \begin{pmatrix} \frac{T^2}{2} \\ T \end{pmatrix}$$

and these correspond to the equations

$$x(k+1) = x(k) + T\dot{x}(k) - \frac{g}{2}T^2$$

$$\dot{x}(k+1) = \dot{x}(k) - gT$$

The parameter T is the update period. The force term produces the acceleration. These are the usual kinematic equations for an accelerating (falling) body.

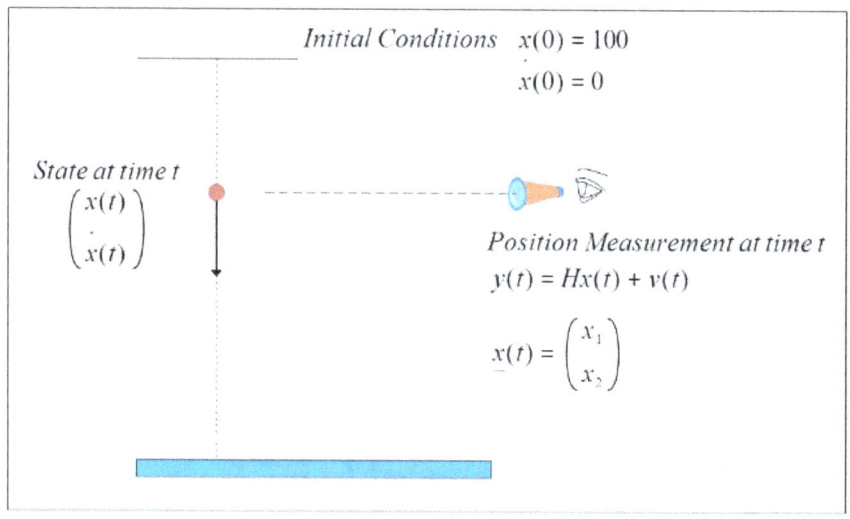

Figure 34 – Free Fall

13.6.2 Results

A sample of the output data are illustrated in Figure 35. The estimated position is seen to track the measured position. The position covariance is initialised to a large value and quickly falls as the estimate converges to an accurate value.

Kalman Filter

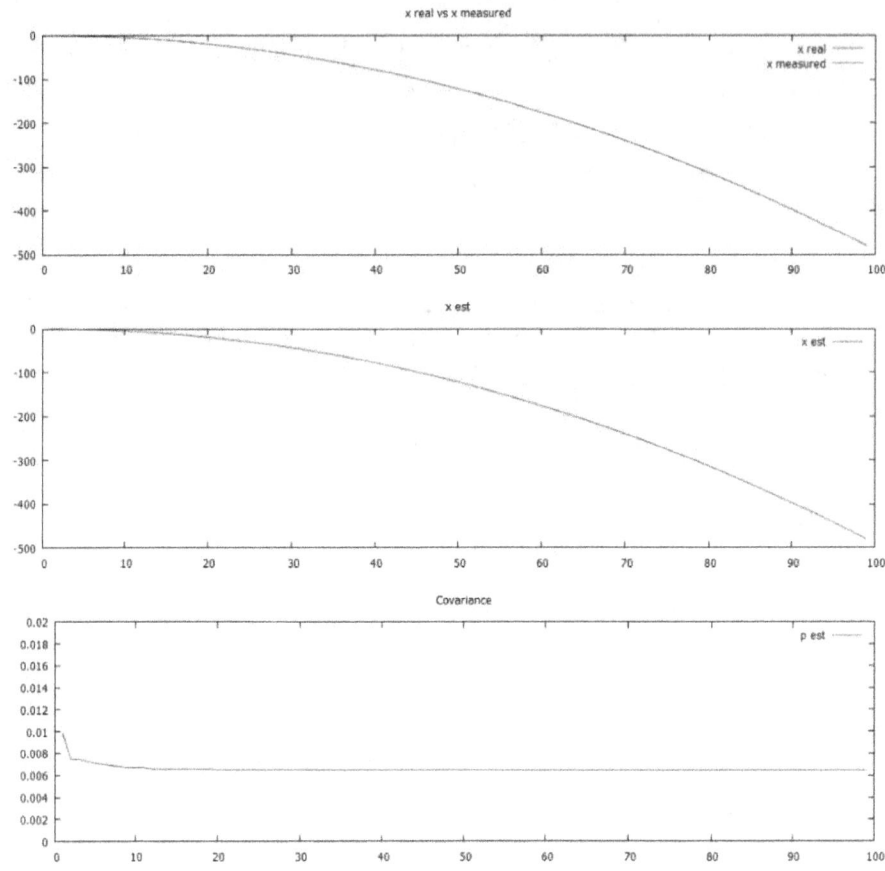

Figure 35 – Falling Body Results

13.6.3 Code

The code presented below corresponds to the above example. It consists only of initialisation, a main loop, and two file handling functions. In contrast to the previous example, this system is a vector Kalman filter and the linear algebra is performed using the GNU Scientific Library (GSL).

Kalman Filter

```c
/*
====================================================
 Name        : FallingBody.c
 Author      : Kalman Filter: Introduction
 Version     : 1.0
 Copyright   : Free to use
 Description : Falling Body Kalman Filter.
====================================================
*/

/*************************************************
       Includes
*************************************************/

#include <stdio.h>
#include <stdlib.h>
#include <math.h>
#include <string.h>

#include <complex.h>
#include <normal.h>
#include <time.h>

#include <gsl/gsl_matrix.h>
#include <gsl/gsl_blas.h>
#include <gsl/gsl_matrix_double.h>
#include <gsl/gsl_linalg.h>

#include <linear_algebra.h>

/*************************************************
       Constants
*************************************************/

#define STDEV_U   0.1
#define STDEV_V   0.1
#define STDEV_W   0.1

#define VAR_U    (STDEV_U * STDEV_U)
#define VAR_V    (STDEV_V * STDEV_V)
#define VAR_W    (STDEV_W * STDEV_W)
```

Kalman Filter

```c
#define U_SIZE  1
#define X_SIZE  2
#define Y_SIZE  1

#define K_MAX   100

/***************************************************
      Typedefs
***************************************************/

/***************************************************
      Types
***************************************************/

/***************************************************
      Globals
***************************************************/

/***************************************************
      Locals
***************************************************/

static int seed = 0;

static double g = 9.8;
static double T = 0.1;

FILE* F_OUT;

char file_out[100] =
"C:\\Users\\andre\\Documents\\Books";

/***************************************************
      Functions
***************************************************/

void open_file(void)
{
  strcat(file_out,
"\\Plot\\FallingBody_KF.dat");
```

Kalman Filter

```c
    F_OUT = fopen(file_out, "w");
}

void close_file(void)
{
      fclose(F_OUT);
}

/****************************************************
      Main
****************************************************/

int main(void)
{
  puts("Falling Body");

  open_file();

  fprintf(F_OUT, "k, x real, y, x_est,
          p_est\n");

  seed = time(0);

  gsl_matrix* PHI =
    gsl_matrix_alloc(X_SIZE, X_SIZE);
  gsl_matrix_set(PHI, 0, 0, 1.0);
  gsl_matrix_set(PHI, 0, 1, T);
  gsl_matrix_set(PHI, 1, 0, 0.0);
  gsl_matrix_set(PHI, 1, 1, 1.0);
  print_matrix(PHI, "PHI INIT", __func__);

  gsl_matrix* B =
    gsl_matrix_alloc(X_SIZE, U_SIZE);
  gsl_matrix_set(B, 0, 0, T * T / 2.0);
  gsl_matrix_set(B, 1, 0, T);

  gsl_matrix* H   =
    gsl_matrix_alloc(Y_SIZE, X_SIZE);
  gsl_matrix_set(H, 0, 0, 1.0);
  gsl_matrix_set(H, 0, 1, 0.0);
```

Kalman Filter

```
gsl_vector* u    = gsl_vector_alloc(U_SIZE);
gsl_vector_set(u, 0, -g);

gsl_matrix* Pest  =
   gsl_matrix_alloc(X_SIZE, X_SIZE);
gsl_matrix_set(Pest, 0, 0, 1.1);
gsl_matrix_set(Pest, 0, 1, 0.0);
gsl_matrix_set(Pest, 1, 0, 0.0);
gsl_matrix_set(Pest, 1, 1, 1.0);

gsl_matrix* Ppred =
   gsl_matrix_alloc(X_SIZE, X_SIZE);
gsl_matrix_set(Ppred, 0, 0, 1.0);
gsl_matrix_set(Ppred, 0, 1, 0.0);
gsl_matrix_set(Ppred, 1, 0, 0.0);
gsl_matrix_set(Ppred, 1, 1, 1.0);

gsl_matrix* Q    =
   gsl_matrix_alloc(X_SIZE, X_SIZE);
gsl_matrix_set(Q, 0, 0, VAR_W);
gsl_matrix_set(Q, 0, 1, 0.0);
gsl_matrix_set(Q, 1, 0, 0.0);
gsl_matrix_set(Q, 1, 1, VAR_W);

gsl_matrix* K    =
   gsl_matrix_alloc(X_SIZE, Y_SIZE);
gsl_matrix_set_all(K, 0.0);

gsl_matrix* R    =
    gsl_matrix_alloc(Y_SIZE, Y_SIZE);
gsl_matrix_set_all(R, VAR_V);

gsl_matrix* D    =
   gsl_matrix_alloc(Y_SIZE, Y_SIZE);
gsl_matrix_set_all(D, 0.0);

gsl_matrix* Dinv   =
   gsl_matrix_alloc(Y_SIZE, Y_SIZE);
gsl_matrix_set_all(Dinv, 0.0);
```

Kalman Filter

```c
gsl_vector* Xest = gsl_vector_alloc(X_SIZE);
gsl_vector* Xpred = gsl_vector_alloc(X_SIZE);
gsl_vector* X = gsl_vector_alloc(X_SIZE);
gsl_vector* Xold = gsl_vector_alloc(X_SIZE);

/* Initialise */
gsl_vector_set(Xold, 0, 0.0);
gsl_vector_set(Xold, 1, 0.0);

gsl_vector_set(Xest, 0, 0.0);
gsl_vector_set(Xest, 1, 0.0);

double x = 0.0;
double x_dot = 0.0;
double t = 0.0;

//float innov[K_MAX] = {0.0};

for (int k = 1; k < K_MAX; k++)
{
    t = t + T;
    printf("k %d, t %f\n", k, t);

    /* Noise */
    float u = r4_normal_01(&seed) * STDEV_U;
    float w = r4_normal_01(&seed) * STDEV_W;
    float v = r4_normal_01(&seed) * STDEV_V;

    /* Actual system --------------------*/

    x = x + x_dot * T - 0.5 * g * T * T;
    x_dot = x_dot - g * T;

    printf("Actual: %f, %f\n", x, x_dot);

    /* x = phi * x_old + Bu + w; */
    gsl_blas_dgemv(
    CblasNoTrans, 1.0, PHI, Xold, 0.0, X);
    gsl_vector* uTemp =
        gsl_vector_alloc(X_SIZE);
    gsl_vector* u_vec =
```

Kalman Filter

```
        gsl_vector_alloc(U_SIZE);
gsl_vector_set(u_vec, 0, -g);
gsl_blas_dgemv(
CblasNoTrans, 1.0, B, u_vec, 0.0, uTemp);
gsl_vector_add(X, uTemp);
gsl_vector* w_temp =
        gsl_vector_alloc(X_SIZE);
gsl_vector_set(w_temp, 0, w);
gsl_vector_set(w_temp, 1, u);
print_vector(Xold, "Xold", __func__);
print_vector(X, "X", __func__);

/* Measurement */
/* y = x + v */
gsl_vector* v_temp =
        gsl_vector_alloc(Y_SIZE);
gsl_vector_set(v_temp, 0, v);
gsl_vector* y =
        gsl_vector_alloc(Y_SIZE);
gsl_vector_set(y, 0, x);
gsl_vector_add(y, v_temp);
print_vector(y, "y", __func__);

/* Prediction------------------------*/

/* x_pred = phi * x_est + Bu; */
gsl_blas_dgemv(
CblasNoTrans,
1.0, PHI, Xest, 0.0, Xpred);
gsl_vector_add(Xpred, uTemp);
print_vector(Xpred, "Xpred", __func__);

/* p_pred = phi * p_est * phi' + Q; */
gsl_matrix* Temp   =
        gsl_matrix_alloc(X_SIZE, X_SIZE);
gsl_blas_dgemm(
  CblasNoTrans, CblasNoTrans,
  1.0, PHI, Pest, 0.0, Temp);
gsl_blas_dgemm(
CblasNoTrans, CblasTrans,
1.0, Temp, PHI, 0.0, Ppred);
```

Kalman Filter

```c
gsl_matrix_add(Ppred, Q);

/* Kalman Gain ----------------------*/

/* D = (H * p_pred * H' + R); */
gsl_matrix* Dtemp =
   gsl_matrix_alloc(Y_SIZE, X_SIZE);
gsl_matrix* Rtemp =
   gsl_matrix_alloc(Y_SIZE, Y_SIZE);
gsl_blas_dgemm(
 CblasNoTrans, CblasNoTrans,
 1.0, H, Ppred, 0.0, Dtemp);
gsl_blas_dgemm(
 CblasNoTrans, CblasTrans,
 1.0, Dtemp, H, 0.0, Rtemp);
gsl_matrix_add(Rtemp, R);
Dinv = matrix_inverse(Rtemp, Y_SIZE);

/* K = p_pred * H' / D; */
print_matrix(Ppred, "Ppred", __func__);
gsl_matrix* Ktemp =
   gsl_matrix_alloc(X_SIZE, Y_SIZE);
gsl_blas_dgemm(
 CblasNoTrans, CblasTrans,
 1.0, Ppred, H, 0.0, Ktemp);
gsl_blas_dgemm(
 CblasNoTrans, CblasNoTrans,
 1.0, Ktemp, Dinv, 0.0, K);
print_matrix(K, "K", __func__);

/* Update ---------------------------*/

/*x_est = x_pred+K * (y-H * x_pred); */
gsl_vector* Ypred =
   gsl_vector_alloc(Y_SIZE);
gsl_vector* Xtemp =
   gsl_vector_alloc(X_SIZE);
gsl_vector* Innov =
   gsl_vector_alloc(Y_SIZE);
gsl_vector_memcpy(Innov, y);
print_vector(y, "y", __func__);
```

Kalman Filter

```
gsl_blas_dgemv(
CblasNoTrans,
1.0, H, Xpred, 0.0, Ypred);
print_vector(Ypred, "Ypred", __func__);
gsl_vector_sub(Innov, Ypred);
print_vector(Innov, "Innov", __func__);
gsl_blas_dgemv(
 CblasNoTrans,
1.0, K, Innov, 0.0, Xtemp);
print_vector(Xpred, "Xpred", __func__);
print_vector(Xtemp, "Xtemp", __func__);
gsl_vector_add(Xpred, Xtemp);
gsl_vector_memcpy(Xest, Xpred);
printf("Actual: %f,%f\n", x, x_dot);
print_vector(Xest, "Xest", __func__);
print_vector(y, "y", __func__);

/* p_est = p_pred - K * H * p_pred; */
gsl_matrix* Ktemp1  =
    gsl_matrix_alloc(X_SIZE, X_SIZE);
gsl_matrix* Ptemp1  =
    gsl_matrix_alloc(X_SIZE, X_SIZE);
gsl_blas_dgemm(
 CblasNoTrans, CblasNoTrans,
1.0, K, H, 0.0, Ktemp1);
gsl_blas_dgemm(
 CblasNoTrans, CblasNoTrans,
1.0, Ktemp1, Ppred, 0.0, Ptemp1);
gsl_matrix_sub(Ppred, Ptemp1);
gsl_matrix_memcpy(Pest, Ppred);
print_matrix(Pest, "Pest", __func__);

/* x_old = x; */
gsl_vector_memcpy(Xold, X);

/* Output results */
fprintf(F_OUT, "%d", k);
fprintf(F_OUT, " %6.2f", x);
fprintf(F_OUT, " %6.2f",
        gsl_vector_get(y, 0));
```

Kalman Filter

```
        fprintf(F_OUT, " %6.2f",
                gsl_vector_get(Xest, 0));
        fprintf(F_OUT, " %6.4f\n",
                gsl_matrix_get(Pest, 0, 0));
    }

    close_file();

    return EXIT_SUCCESS;
}
```

13.7 Radar System

A radar system is simulated in this example. This is a standard example and is based on the presentation in [3]. This in turn is based on the original paper [4].

13.7.1 Description

The system is represented in plane polar coordinates and estimates are computed for the range, range rate, bearing, and bearing rate of the target. Measurements, with additive noise, are made of the range and bearing. The target trajectory is illustrated in Figure 36. The initial conditions are as shown in the diagram.

Kalman Filter

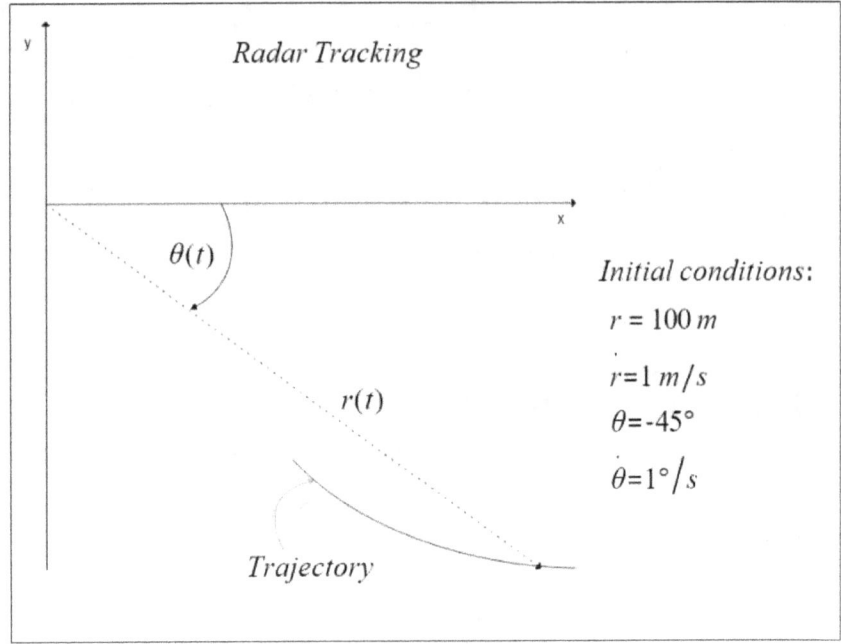

Figure 36 – Radar Target Trajectory

13.7.2 Results

Some typical results from the radar simulation are presented in Figure 35. The r coordinate estimate tracks the real value accurately. The theta coordinate is also accurate but is noisier than the radial coordinate. The covariance matrix entry corresponding to the radial coordinate, starts high and quickly falls to the expected value for the error covariance (variance) of the radial coordinate.

Kalman Filter

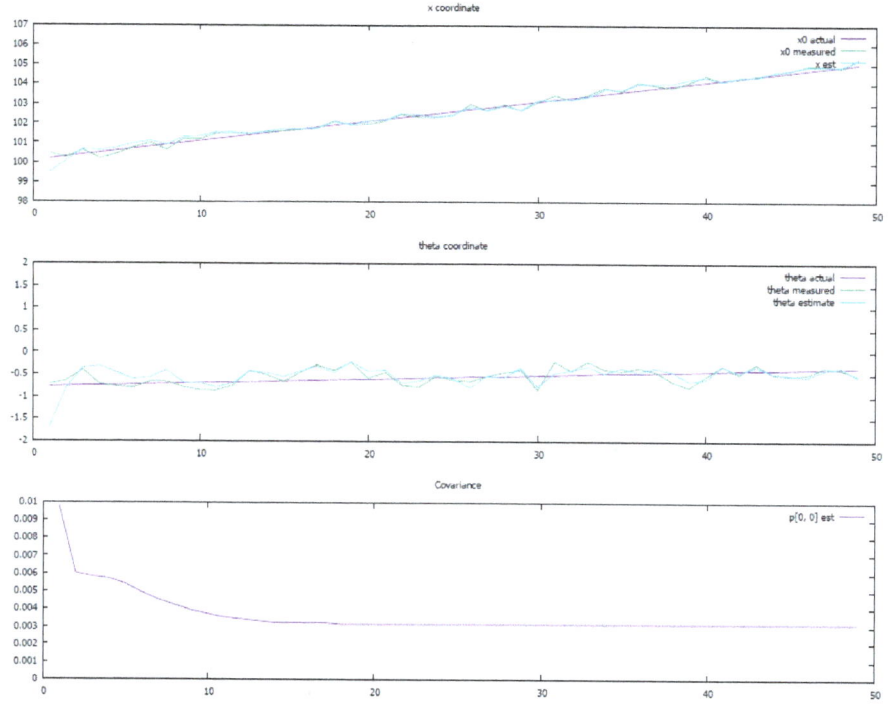

Figure 37 - Radar Results

13.7.3 Code

The code presented below corresponds to the above example. It consists only of initialisation, a main loop, and two file handling functions. This system is a vector Kalman filter and the linear algebra is performed using the GNU Scientific Library (GSL).

Kalman Filter

```c
/*
=====================================================
    Name        : Radar.c
    Author      : Kalman Filter: Introduction
    Version     : 1.0
    Copyright   : Free to use
    Description : Radar Kalman Filter.
=====================================================*/

/************************************************
      Includes
************************************************/

#include <stdio.h>
#include <stdlib.h>
#include <math.h>
#include <string.h>

#include <complex.h>
#include <normal.h>
#include <time.h>

#include <gsl/gsl_matrix.h>
#include <gsl/gsl_blas.h>
#include <gsl/gsl_matrix_double.h>
#include <gsl/gsl_linalg.h>

#include <linear_algebra.h>

/************************************************
      Constants
************************************************/

#define PI 3.14159

#define STDEV_U    0.1
#define STDEV_V    0.1
#define STDEV_W    0.1

#define VAR_U    (STDEV_U * STDEV_U)
#define VAR_V    (STDEV_V * STDEV_V)
```

Kalman Filter

```c
#define VAR_W   (STDEV_W * STDEV_W)

#define X_SIZE 4
#define Y_SIZE 2

#define K_MAX   50

#define R_DOT    1.0
#define THETA_DOT  (5.0 * PI / 180.0)

/************************************************
    Typedefs
************************************************/

/************************************************
    Types
************************************************/

/************************************************
    Globals
************************************************/

/************************************************
    Locals
************************************************/

static int seed = 0;
static double T = 0.1;

FILE* F_OUT;

char file_out[100] =
"C:\\Users\\andre\\Documents\\Books";

/************************************************
    Functions
************************************************/

void open_file(void)
{
  strcat(file_out, "\\Plot\\Radar_KF.dat");
```

Kalman Filter

```c
    F_OUT = fopen(file_out, "w");
}

void close_file(void)
{
    fclose(F_OUT);
}

/***********************************************
      Main
***********************************************/

int main(void)
{
  puts("Bias Estimation");

  open_file();

  fprintf(F_OUT, "k, x actual, x meas, x est,
    theta actual, theta meas, theta est,
    p_est\n");

  seed = time(0);

  gsl_matrix* PHI =
      gsl_matrix_alloc(X_SIZE, X_SIZE);
  gsl_matrix_set(PHI, 0, 0, 1.0); // r
  gsl_matrix_set(PHI, 0, 1, T);
  gsl_matrix_set(PHI, 0, 2, 0.0);
  gsl_matrix_set(PHI, 0, 3, 0.0);
  gsl_matrix_set(PHI, 1, 0, 0.0); // r_dot
  gsl_matrix_set(PHI, 1, 1, 1.0);
  gsl_matrix_set(PHI, 1, 2, 0.0);
  gsl_matrix_set(PHI, 1, 3, 0.0);
  gsl_matrix_set(PHI, 2, 0, 0.0); // theta
  gsl_matrix_set(PHI, 2, 1, 0.0);
  gsl_matrix_set(PHI, 2, 2, 1.0);
  gsl_matrix_set(PHI, 2, 3, T);
  gsl_matrix_set(PHI, 3, 0, 0.0); // theta_dot
  gsl_matrix_set(PHI, 3, 1, 0.0);
```

Kalman Filter

```
gsl_matrix_set(PHI, 3, 2, 0.0);
gsl_matrix_set(PHI, 3, 3, 1.0);

gsl_matrix* H     =
    gsl_matrix_alloc(Y_SIZE, X_SIZE);
gsl_matrix_set(H, 0, 0, 1.0);
gsl_matrix_set(H, 0, 1, 0.0);
gsl_matrix_set(H, 0, 2, 0.0);
gsl_matrix_set(H, 0, 3, 0.0);
gsl_matrix_set(H, 1, 0, 0.0);
gsl_matrix_set(H, 1, 1, 0.0);
gsl_matrix_set(H, 1, 2, 1.0);
gsl_matrix_set(H, 1, 3, 0.0);

gsl_matrix* Pest  =
    gsl_matrix_alloc(X_SIZE, X_SIZE);
gsl_matrix_set(Pest, 0, 0, 1.0);
gsl_matrix_set(Pest, 0, 1, 0.0);
gsl_matrix_set(Pest, 0, 2, 0.0);
gsl_matrix_set(Pest, 0, 3, 0.0);
gsl_matrix_set(Pest, 1, 0, 0.0);
gsl_matrix_set(Pest, 1, 1, 1.0);
gsl_matrix_set(Pest, 1, 2, 0.0);
gsl_matrix_set(Pest, 1, 3, 0.0);
gsl_matrix_set(Pest, 2, 0, 0.0);
gsl_matrix_set(Pest, 2, 1, 0.0);
gsl_matrix_set(Pest, 2, 2, 1.0);
gsl_matrix_set(Pest, 2, 3, 0.0);
gsl_matrix_set(Pest, 3, 0, 0.0);
gsl_matrix_set(Pest, 3, 1, 0.0);
gsl_matrix_set(Pest, 3, 2, 0.0);
gsl_matrix_set(Pest, 3, 3, 1.0);

gsl_matrix* Ppred =
    gsl_matrix_alloc(X_SIZE, X_SIZE);
gsl_matrix_set(Ppred, 0, 0, 1.0);
gsl_matrix_set(Ppred, 0, 1, 0.0);
gsl_matrix_set(Ppred, 0, 2, 0.0);
gsl_matrix_set(Ppred, 0, 3, 0.0);
gsl_matrix_set(Ppred, 1, 0, 0.0);
gsl_matrix_set(Ppred, 1, 1, 1.0);
```

Kalman Filter

```
gsl_matrix_set(Ppred, 1, 2, 0.0);
gsl_matrix_set(Ppred, 1, 3, 0.0);
gsl_matrix_set(Ppred, 2, 0, 0.0);
gsl_matrix_set(Ppred, 2, 1, 0.0);
gsl_matrix_set(Ppred, 2, 2, 1.0);
gsl_matrix_set(Ppred, 2, 3, 0.0);
gsl_matrix_set(Ppred, 3, 0, 0.0);
gsl_matrix_set(Ppred, 3, 1, 0.0);
gsl_matrix_set(Ppred, 3, 2, 0.0);
gsl_matrix_set(Ppred, 3, 3, 1.0);

gsl_matrix* Q   =
    gsl_matrix_alloc(X_SIZE, X_SIZE);
gsl_matrix_set(Q, 0, 0, 0.0);
gsl_matrix_set(Q, 0, 1, 0.0);
gsl_matrix_set(Q, 0, 2, 0.0);
gsl_matrix_set(Q, 0, 3, 0.0);
gsl_matrix_set(Q, 1, 0, 0.0);
gsl_matrix_set(Q, 1, 1, VAR_W);
gsl_matrix_set(Q, 1, 2, 0.0);
gsl_matrix_set(Q, 1, 3, 0.0);
gsl_matrix_set(Q, 2, 0, 0.0);
gsl_matrix_set(Q, 2, 1, 0.0);
gsl_matrix_set(Q, 2, 2, 0.0);
gsl_matrix_set(Q, 2, 3, 0.0);
gsl_matrix_set(Q, 3, 0, 0.0);
gsl_matrix_set(Q, 3, 1, 0.0);
gsl_matrix_set(Q, 3, 2, 0.0);
gsl_matrix_set(Q, 3, 3, VAR_W);

gsl_matrix* K   =
    gsl_matrix_alloc(X_SIZE, Y_SIZE);
gsl_matrix_set_all(K, 0.0);

gsl_matrix* R   =
    gsl_matrix_alloc(Y_SIZE, Y_SIZE);
gsl_matrix_set_all(R, VAR_V);

gsl_matrix* D   =
    gsl_matrix_alloc(Y_SIZE, Y_SIZE);
gsl_matrix_set_all(D, 0.0);
```

Kalman Filter

```c
gsl_matrix* Dinv   =
    gsl_matrix_alloc(Y_SIZE, Y_SIZE);
gsl_matrix_set_all(Dinv, 0.0);

gsl_vector* Xest = gsl_vector_alloc(X_SIZE);
gsl_vector* Xpred = gsl_vector_alloc(X_SIZE);
gsl_vector* X = gsl_vector_alloc(X_SIZE);
gsl_vector* Xold = gsl_vector_alloc(X_SIZE);

/* Initialise */
gsl_vector_set(Xold, 0, 0.0);
gsl_vector_set(Xold, 1, 0.0);
gsl_vector_set(Xold, 2, 0.0);
gsl_vector_set(Xold, 3, 0.0);

gsl_vector_set(Xest, 0, 0.0);
gsl_vector_set(Xest, 1, 0.0);
gsl_vector_set(Xest, 2, 0.0);
gsl_vector_set(Xest, 3, 0.0);

double r = 100.0;
double r_dot = 1.0;
double theta = -45.0 * PI / 180.0;
double theta_dot = 1.0 * PI / 180.0;

double t = 0.0;

/* Initialise with two measurements */
/* theta = theta + v */
float u = r4_normal_01(&seed) * STDEV_U;
float w = r4_normal_01(&seed) * STDEV_W;
float v1 = r4_normal_01(&seed) * STDEV_V;
float v2 = r4_normal_01(&seed) * STDEV_V;

double r_meas = r + v1;
double theta_meas = theta + v2;

double r_temp = r_meas;
double theta_temp = theta_meas;
```

Kalman Filter

```
// Update
r_temp = r;
double r_dot_temp = r_dot;
theta_temp = theta;
double theta_dot_temp = theta_dot;

r = r_temp + r_dot * T;
r_dot = r_dot_temp;
theta = theta_temp + theta_dot * T;
theta_dot = theta_dot_temp;

/* r = r + v */
/* theta = theta + v */
u = r4_normal_01(&seed) * STDEV_U;
w = r4_normal_01(&seed) * STDEV_W;
v1 = r4_normal_01(&seed) * STDEV_V;
v2 = r4_normal_01(&seed) * STDEV_V;

r_meas = r + v1;
double r_dot_meas = (r_meas - r_temp) / T;
theta_meas = theta + v2;
double theta_dot_meas =
    (theta_meas - theta_temp) / T;

gsl_vector_set(Xold, 0, r_meas);
gsl_vector_set(Xold, 1, r_dot_meas);
gsl_vector_set(Xold, 2, theta_meas);
gsl_vector_set(Xold, 03, theta_dot_meas);

/* Simulation loop */
for (int k = 1; k < K_MAX; k++)
{
    t = t + T;

    /* Noise */
    float u = r4_normal_01(&seed) * STDEV_U;
    float w1 =
        r4_normal_01(&seed) * STDEV_W;
    float w2 =
        r4_normal_01(&seed) * STDEV_W;
    float w3 =
```

Kalman Filter

```c
        r4_normal_01(&seed) * STDEV_W;
    float w4 =
        r4_normal_01(&seed) * STDEV_W;
    float v1 =
        r4_normal_01(&seed) * STDEV_V;
    float v2 =
        r4_normal_01(&seed) * STDEV_V;

    // Actual system ------------------------

    // System
    //double r_temp = r;
    double r_dot_temp = R_DOT;
    //double theta_temp = theta;
    double theta_dot_temp = THETA_DOT;
    r = r + r_dot * T;
    r_dot = r_dot_temp;
    theta = theta + theta_dot * T;
    theta_dot = theta_dot_temp;

    /* x = phi * x_old + w; */
    gsl_blas_dgemv(
    CblasNoTrans,
    1.0, PHI, Xold, 0.0, X);
    gsl_vector* w_temp =
        gsl_vector_alloc(X_SIZE);
    gsl_vector_set(w_temp, 0, w1);
    gsl_vector_set(w_temp, 1, w2);
    gsl_vector_set(w_temp, 2, w3);
    gsl_vector_set(w_temp, 3, w4);
    gsl_vector_add(X, w_temp);

    // Measurement
    /* r = r + v */
    /* theta = theta + v */
    double r_meas = r + v1;
    double theta_meas = theta + v2;

    gsl_vector* v_temp =
        gsl_vector_alloc(Y_SIZE);
    gsl_vector_set(v_temp, 0, v1);
```

Kalman Filter

```
gsl_vector_set(v_temp, 1, v2);
gsl_vector* y =
  gsl_vector_alloc(Y_SIZE);
gsl_vector_set(y, 0, r_meas);
gsl_vector_set(y, 1, theta_meas);
gsl_vector_add(y, v_temp);

// Prediction--------------------------

/* x_pred = phi * x_est; */
gsl_blas_dgemv(
CblasNoTrans,
1.0, PHI, Xest, 0.0, Xpred);

//p_pred = phi * p_est * phi' + Q;
gsl_matrix* Temp   =
  gsl_matrix_alloc(X_SIZE, X_SIZE);
gsl_blas_dgemm(
CblasNoTrans, CblasNoTrans,
1.0, PHI, Pest, 0.0, Temp);
gsl_blas_dgemm(CblasNoTrans, CblasTrans,
1.0, Temp, PHI, 0.0, Ppred);
gsl_matrix_add(Ppred, Q);

// Kalman Gain ------------------------

/* D =  (H * p_pred * H' + R); */
gsl_matrix* Dtemp   =
  gsl_matrix_alloc(Y_SIZE, X_SIZE);
gsl_matrix* Rtemp   =
  gsl_matrix_alloc(Y_SIZE, Y_SIZE);
gsl_blas_dgemm(
CblasNoTrans, CblasNoTrans,
1.0, H, Ppred, 0.0, Dtemp);
gsl_blas_dgemm(
CblasNoTrans, CblasTrans,
1.0, Dtemp, H, 0.0, Rtemp);
gsl_matrix_add(Rtemp, R);
Dinv = matrix_inverse(Rtemp, Y_SIZE);

/* K = p_pred * H' / D; */
```

Kalman Filter

```
gsl_matrix* Ktemp =
   gsl_matrix_alloc(X_SIZE, Y_SIZE);
gsl_blas_dgemm(
 CblasNoTrans, CblasTrans,
 1.0, Ppred, H, 0.0, Ktemp);
gsl_blas_dgemm(
 CblasNoTrans, CblasNoTrans,
 1.0, Ktemp, Dinv, 0.0, K);

// Update ------------------------------

/* x_est = x_pred+K * (y-H * x_pred); */
gsl_vector* Ypred   =
   gsl_vector_alloc(Y_SIZE);
gsl_vector* Xtemp   =
   gsl_vector_alloc(X_SIZE);
gsl_vector* Innov   =
   gsl_vector_alloc(Y_SIZE);
gsl_vector_memcpy(Innov, y);
gsl_blas_dgemv(
 CblasNoTrans,
 1.0, H, Xpred, 0.0, Ypred);
gsl_vector_sub(Innov, Ypred);
gsl_blas_dgemv(
 CblasNoTrans,
 1.0, K, Innov, 0.0, Xtemp);
gsl_vector_add(Xpred, Xtemp);
gsl_vector_memcpy(Xest, Xpred);

/* p_est = p_pred - K * H * p_pred; */
gsl_matrix* Ktemp1  =
   gsl_matrix_alloc(X_SIZE, X_SIZE);
gsl_matrix* Ptemp1  =
   gsl_matrix_alloc(X_SIZE, X_SIZE);
gsl_blas_dgemm(
 CblasNoTrans, CblasNoTrans,
 1.0, K, H, 0.0, Ktemp1);
gsl_blas_dgemm(
 CblasNoTrans, CblasNoTrans,
 1.0, Ktemp1, Ppred, 0.0, Ptemp1);
gsl_matrix_sub(Ppred, Ptemp1);
```

Kalman Filter

```c
            gsl_matrix_memcpy(Pest, Ppred);

            /* x_old = x; */
            gsl_vector_memcpy(Xold, X);

            // Results -----------------------------

            /* Output results */
            fprintf(F_OUT, "%d", k);  // k 1

            fprintf(F_OUT, "%6.2f", r);  // x actual 2
            fprintf(F_OUT, " %6.2f",
                gsl_vector_get(y, 0));  // x meas 3
            fprintf(F_OUT, " %6.2f",
                gsl_vector_get(Xest, 0));  // x est 4

            fprintf(F_OUT, " %6.2f", theta);
            // theta actual 5
            fprintf(F_OUT, " %6.2f",
                gsl_vector_get(y, 1));
            // theta meas 6
            fprintf(F_OUT, " %6.2f",
                gsl_vector_get(Xest, 2));
            // theta est 7

            fprintf(F_OUT, " %6.4f\n",
                gsl_matrix_get(Pest, 0, 0));
            // P[0, 0] est 8
    }

    puts("End");

    return EXIT_SUCCESS;
}
```

13.8 Utility Functions

The following external functions are called from the code.

Kalman Filter

13.8.1 Gaussian Random Noise

The following open-source code (GNU LGPL) was used to generate Gaussian pseudorandom numbers to simulate plant and measurement noise.

```
/*****************************************************
  Purpose:
    R4_NORMAL_01 returns a unit pseudonormal R4.
  Discussion:
    The standard normal probability distribution
    function (PDF) has mean 0 and standard
    deviation 1.
    The Box-Muller method is used, which is
    efficient, but generates two values at a
    time.
  Licensing:
    This code is distributed under the
    GNU LGPL license.
  Modified:
    05 June 2013
  Author:
    John Burkardt

  Parameters:
    Input/output, int *SEED,
    a seed for the random number generator.

    Output, float R4_NORMAL_01,
    a normally distributed random value.
*****************************************************/

float r4_normal_01 (int *seed)
{
  float r1;
  float r2;
  const double r4_pi = 3.141592653589793;
  float x;

  r1 = r4_uniform_01(seed);
  r2 = r4_uniform_01(seed);
```

Kalman Filter

```
    x =
    sqrt(-2.0 * log(r1)) * cos(2.0 * r4_pi * r2);

    return x;
}
```

13.8.2   Matrix Inversion

The GSL is used to compute the inverse of a square matrix by LU decomposition. The code is presented below. The decomposition uses a pivoting algorithm and the permutation matrix is used to keep track of the swapping of rows and columns during this process.

```
/*******************************************
    Function: matrix_inverse

    Print the inverse of a matrix.

    In: gsl_matrix* m Matrix to invert
    In: int size Size of the matrix

    Return: gsl_matrix* pointer to inverted
    matrix
*******************************************/

gsl_matrix* matrix_inverse(gsl_matrix* M, int size)
{
  gsl_permutation *p =
     gsl_permutation_alloc(size);
  int s;

  /* Compute LU decomposition of this matrix */
  gsl_linalg_LU_decomp(M, p, &s);

  /* Compute inverse of the LU decomposition */
  gsl_matrix *inv =
     gsl_matrix_alloc(size, size);
```

Kalman Filter

```
  gsl_linalg_LU_invert(M, p, inv);

  gsl_permutation_free(p);

  return inv;
}
```

14 Further Reading

The sections below discuss further steps in the study of the Kalman filter.

14.1 Extended Kalman Filter

For a linear system as discussed earlier, the system dynamics, the process model and the measurement model, can be represented by matrices and vectors. The dynamic system and the measurement functions are linear with respect to the state variables. The Extended Kalman filter (EKF) provides a means of applying the Kalman filter to non-linear systems. This non-linear version of the Kalman filter is realised by using a Taylor series to linearise the model about the current estimate. The EKF is not an optimal filter because it is implemented in terms of a set of approximations. The resulting algorithm is very similar to that of the linear Kalman filter. A typical application is target tracking where polar coordinates used for observations must be converted to the cartesian coordinates of the state vector.

14.2 Square Root Kalman Filter

The Kalman filter can suffer from numerical instability. If the process noise is small and round-off error causes a positive eigenvalue to be computed as a negative value then the covariance matrix may no longer be positive semi-definite. The symmetric, positive semi-definite covariance matrix can be factored into the product of a matrix and its transpose as in

$$P = S\,S^T$$

and then, if the updates are made to the square root matrix (S), the covariance matrix formed from the above product is guaranteed to be positive semi-definite and symmetric. The factorisation of the covariance matrix is achieved using the Cholesky factorisation algorithm. The square root of a symmetric matrix is not unique and so other forms of factorisation are also possible.

The first square root algorithm was the Potter square root filter. It

Kalman Filter

was for the simplest case of uncorrelated scalar observations with no process noise. The concept involves Cholesky factorisation of the covariance matrix and the update of the Cholesky factors. It proceeds as follows.

The covariance matrix is factored according to

$$P(k|k-1) = S(k|k-1)S^T(k|k-1)$$
$$P(k|k) = S(k|k)S^T(k|k)$$

The observation update is then

$$P(k|k) = P(k|k-1) - K(k)HP(k|k-1)$$

and this becomes

$$S(k|k)S^T(k|k) = S(k|k-1)S^T(k|k-1) -$$
$$S(k|k-1)S^T(k|k-1)H^T[X(k|k-1)]^{-1}HS(k|k-1)S^T(k|k-1)$$

where:

$$X(k|k-1) = HS(k|k-1)S^T(k|k-1)H^T + R$$

and this can be written as

$$S(k|k)S^T(k|k) =$$
$$S(k|k-1)S^T(k|k-1) - S(k|k-1)V[V^TV + R]^{-1}V^TS^T(k|k-1)$$
$$= S(k|k-1)[1 - V(V^TV + R)^{-1}V^T]S^T(k|k-1)$$

The matrix $1 - V(V^TV + R)^{-1}V^T$ is symmetric. This is then factored as

$$WW^T = 1 - V(V^TV + R)^{-1}V^T$$

and the update of $S(k|k)S^T(k|k)$ becomes

$$S(k|k)S^T(k|k) = S(k|k-1)WW^TS^T(k|k-1)$$

and then the update of $S(k|k)$ is

$$S(k|k) = S(k|k-1)W$$

The processing then continues in terms of S rather than P. The Cholesky factors are better conditioned for matrix operations than the covariance matrices and this leads to a more robust algorithm in the presence of round-off errors.

Kalman Filter

14.3 Noise Models

The Kalman filter is the optimal linear filter even in the presence of non-Gaussian noise. However, it is assumed that process noise and measurement noise are uncorrelated. If this is not the case then updated Kalman filter algorithms have been developed accordingly. Additionally, algorithms have been developed for the detection and modelling of coloured noise. These use autoregressive models to provide a noise shaping filter which is then incorporated into the Kalman filter algorithm.

14.4 Adaptive Kalman Filter

The Kalman filter may be modified to allow the filter to adapt the measurement model and/or the process model according to the measured data. This can be applied to the problem of a manoeuvring target.

14.5 Data Fusion

The Kalman filter typically uses measurements from multiple sources to track an object. This is an example of sensor fusion. The combined data from a number of sensors provides a more accurate estimate of the state than could be obtained from any of the individual sensors.

14.6 Numerical Methods

As mentioned above, numerical stability can be an issue with the Kalman filter. Numerical methods may be applied to improve the stability of the implementation. These include Cholesky factorisation, UD factorisation, and LU decomposition. These are all methods that are used to maintain a symmetric and positive definite covariance matrix in the presence of round-off errors.

15 Appendices

These appendices discuss in detail various items that were deferred from the main text.

15.1 Matrix Identities

Some useful matrix identities are summarised here.

15.1.1 The Jacobian

Let $\mathbf{y} = \mathbf{f}(\mathbf{x})$ where $\mathbf{y}$ and $\mathbf{x}$ are vectors of order m and n respectively. Each element of $\mathbf{y}$ can be differentiated wrt each element of $\mathbf{x}$ and then we obtain the matrix, known as the Jacobian of the transformation f, given by

$$J = \frac{\partial(f_1,...f_m)}{\partial(x_1,...x_n)} = \begin{pmatrix} \frac{\partial y_1}{\partial x_1} & \cdots & \frac{\partial y_1}{\partial x_n} \\ \vdots & \ddots & \vdots \\ \frac{\partial y_m}{\partial x_1} & \cdots & \frac{\partial y_m}{\partial x_m} \end{pmatrix}$$

When making a change of variables defined by the functions f_i, the Jacobian matrix is square and in this case the determinant of the Jacobian matrix is required to transform the area or volume differentials from one coordinate system to the other.

For example, when making a transformation from Cartesian coordinates to polar coordinates:

$$x = r cos\theta$$
$$y = r sin\theta$$

the area element becomes:

$$dxdy \rightarrow rd\theta dr$$

The Jacobian matrix is

$$J = \begin{pmatrix} cos\theta & -rsin\theta \\ sin\theta & rcos\theta \end{pmatrix} = rcos^2\theta + rsin^2\theta = r$$

$$|J| = r$$

Kalman Filter

The Jacobian determinant is the 'r' term in the transformed area element.

15.1.2 Inversion Relation

Let the matrix M be partitioned into submatrices as follows (where matrices A B C D have consistent dimensions):

$$M = \begin{pmatrix} A & B \\ C & D \end{pmatrix}$$

Then we can express the inverse of M as

$$M^{-1} = \begin{pmatrix} (A - BD^{-1}C)^{-1} & -A^{-1}B(D - CA^{-1}B)^{-1} \\ -D^{-1}C(A - BD^{-1}C)^{-1} & (D - CA^{-1}B)^{-1} \end{pmatrix}$$

By matrix multiplication we can verify this by multiplying MM^{-1} as follows

$$(MM^{-1})_{11} = A(A - BD^{-1}C)^{-1} -$$
$$BD^{-1}C(A - BD^{-1}C)^{-1}$$
$$= (A - BD^{-1}C)(A - BD^{-1}C)^{-1} = 1$$
$$(MM^{-1})_{12} = -AA^{-1}B(D - CA^{-1}B)^{-1} +$$
$$B(D - CA^{-1}B)^{-1}$$
$$= -B(D - CA^{-1}B)^{-1} + B(D - CA^{-1}B)^{-1}$$
$$= 0$$

and similarly for the other two elements.

15.2 Matrix Differentiation

It is often easier to use index notation when dealing with matrices but, in the case of the Kalman filter, matrix notation often is used. This includes the result of differentiation of matrices, vectors and other combinations of matrices and vectors. In this section we discuss some useful matrix identities.

Kalman Filter

15.2.1 Simple Product

Let $y = A x$ where A is constant matrix then,

$$y_i = \sum_{j=1}^{n} a_{ij} x_j$$

Hence, differentiating,

$$\frac{\partial y_i}{\partial x_k} = \sum_{j=1}^{n} a_{ij} \frac{\partial x_j}{\partial x_k} = \sum_{j=1}^{n} a_{ij} \delta_{jk} = a_{ik}$$

$$\frac{\partial y}{\partial x} = A$$

15.2.2 Bilinear Form

Let a scalar be given by

$$s = x^T A x$$

where A is a constant square matrix. Using the Leibnitz rule, we obtain

$$s = \sum_{j=1}^{n} \sum_{k=1}^{n} a_{jk} x_j x_k$$

$$\frac{\partial s}{\partial x_i} = \sum_{j=1}^{n} a_{ij} x_j + \sum_{k=1}^{n} a_{ki} x_k$$

In matrix notation this is

$$\frac{\partial s}{\partial x} = x^T A^T + x^T A = x^T (A^T + A)$$

If A is symmetric (i.e., $A = A^T$) then

$$\frac{\partial s}{\partial x} = 2 x^T A$$

For the case where $x = x(z)$, $A = 1$, then (using the chain rule)

$$\frac{\partial s}{\partial z} = 2 x^T \frac{\partial x}{\partial z}$$

15.2.3 Differentiate Matrix wrt Scalar

If A is a matrix whose elements are functions of a scalar s then

Kalman Filter

$$\frac{\partial A}{\partial s} = \begin{pmatrix} \frac{\partial a_{11}}{\partial s} & \cdots & \frac{\partial a_{1n}}{\partial s} \\ \vdots & \ddots & \vdots \\ \frac{\partial a_{m1}}{\partial s} & \cdots & \frac{\partial a_{mn}}{\partial s} \end{pmatrix}$$

15.2.4 Differentiate Inverse

If A is a non-singular $n \times n$ matrix, whose elements are functions of a scalar s, then differentiate the relation $A^{-1}A = 1$ to yield

$$\frac{\partial A^{-1}}{\partial a} A + A^{-1} \frac{\partial A}{\partial s} = 0$$

Hence

$$\frac{\partial A^{-1}}{\partial a} = -A^{-1} \frac{\partial A}{\partial s} A^{-1}$$

15.3 Curvature

Curvature is introduced as a prelude to the discussion of Fisher information and the Cramer Rao lower bound.

15.3.1 Motivation

The curvature of a coplanar line, at a point P, is measured by finding the 'best' circle that is tangent to the curve at the point P. Some candidates are illustrated in Figure 38. The circle S_1 is the closest fit to the line at the point P.

Kalman Filter

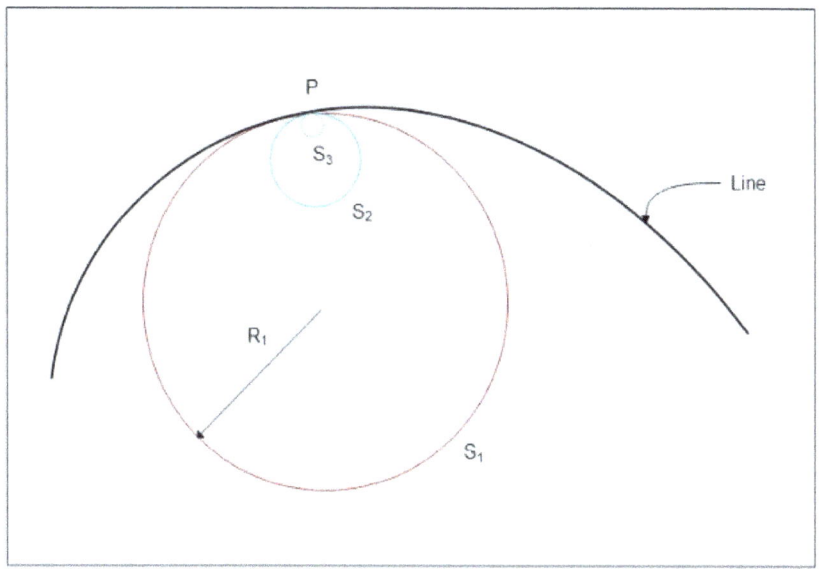

Figure 38 – Curvature of a Line

In Figure 39 it can be seen that the high curvature corresponds to a circle with small radius. The low curvature corresponds to a circle with a large radius. Hence, it is reasonable to assume that a measure of curvature could be an expression of the form

$$\kappa = \frac{1}{R}$$

where:

κ is the standard symbol for curvature

R is the radius of curvature

If a curve has a radius of curvature R, at a point P, then the circle of radius R which has the same tangent at P as the curve, is called the osculating circle to the curve at P. The centre of the osculating circle lies on the concave side of the curve, as in the figures.

For a straight line the curvature is zero and the radius of curvature is infinite.

A curve with constant non-zero curvature κ is a circle of radius $\frac{1}{\kappa}$.

Kalman Filter

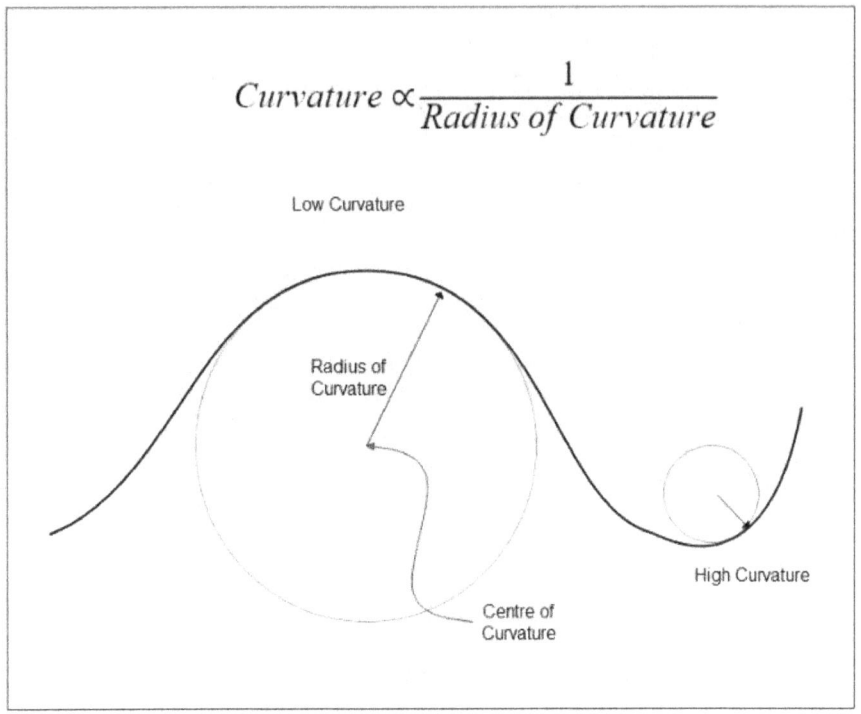

Figure 39 – Curvature and Radius

In Figure 40 it is seen that for a line with high curvature the tangent to the curve changes value quickly whereas for a line with low curvature the tangent to the line changes value slowly. This leads us to postulate the curvature and the rate of change of the gradient are related. The expected relation, for tangent T, is something like

$$\kappa = \frac{dT}{dx} = \frac{d}{dx}\left(\frac{dy}{dx}\right)$$

and so, the curvature could be related to the second derivative of the curve

$$\kappa = \frac{d^2y}{dx^2}$$

This is investigated in more detail in the following section.

Kalman Filter

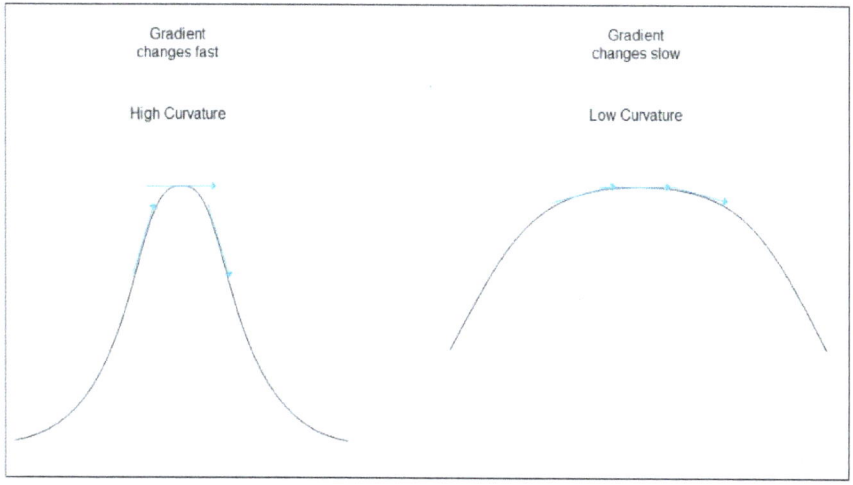

Figure 40 – Change of Gradient

15.3.2   Curvature of a Planar Curve

The rate of change of θ with arc length s is called the curvature of a curve, and is denoted by κ. We have previously motivated the idea of curvature and now we derive an expression for the curvature of a function. The geometry is illustrated in Figure 41.

The curvature measures how rapidly the direction of the tangent line is changing, as a function of position on the curve. It is given by the modulus because the curvature must be a positive value.

$$\kappa = \left| \frac{d\theta}{ds} \right|$$

The gradient is

$$\frac{dy}{dx} = tan\theta$$

and then

Kalman Filter

$$\frac{d}{dx}\left(\frac{dy}{dx}\right) = \frac{d\theta}{dx}\frac{d}{d\theta}\left(\frac{dy}{dx}\right) = \frac{d\theta}{dx}\frac{d}{d\theta}\tan\theta = \sec^2\theta \frac{d\theta}{dx}$$

$$= (1 + \tan^2\theta)\frac{d\theta}{dx} = \left[1 + \left(\frac{dy}{dx}\right)^2\right]\frac{d\theta}{dx}$$

but also

$$\frac{d}{ds}\left(\frac{dy}{dx}\right) = \frac{dx}{ds}\frac{d}{dx}\left(\frac{dy}{dx}\right) = \frac{dx}{ds}\frac{d^2y}{dx^2}$$

The arc length is given by Pythagoras' theorem

$$ds^2 = dx^2 + dy^2$$

$$ds = (dx^2 + dy^2)^{1/2} = dx\left(1 + \left(\frac{dy}{dx}\right)^2\right)^{1/2}$$

$$\frac{dx}{ds} = \left(1 + \left(\frac{dy}{dx}\right)^2\right)^{-1/2}$$

and then

$$\frac{d\theta}{ds} = \frac{d\theta}{dx}\frac{dx}{ds} = \frac{d^2y}{dx^2}\left[1 + \left(\frac{dy}{dx}\right)^2\right]^{-1}\left[1 + \left(\frac{dy}{dx}\right)^2\right]^{-1/2}$$

$$\kappa = \left|\frac{d\theta}{ds}\right| = \frac{\left|\frac{d^2y}{dx^2}\right|}{\left[1 + \left(\frac{dy}{dx}\right)^2\right]^{3/2}}$$

This meaning of this expression is understood from our motivational discussion in 15.3.1.

Kalman Filter

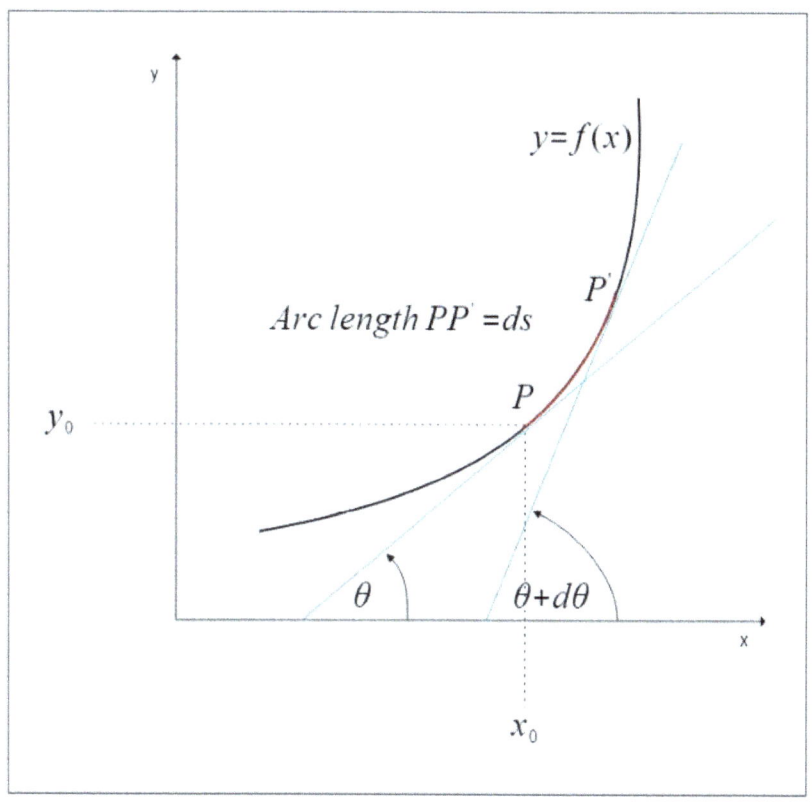

Figure 41 – Curvature of a Line

15.3.3 Example: Curvature of a Parabola

Find the curvature of the parabola $y = \frac{1}{2}x^2$ at the point $x = 0$.
The curvature is given by

$$\kappa = \frac{\left|\frac{d^2y}{dx^2}\right|}{\left[1+\left(\frac{dy}{dx}\right)^2\right]^{3/2}}$$

and for the quoted parabola this is arrived at as follows

$$y = \frac{1}{2}x^2$$

Kalman Filter

$$\frac{dy}{dx} = x$$

$$\frac{d^2y}{dx^2} = 1$$

and therefore

$$\kappa = \frac{1}{(1+x^2)^{3/2}}$$

The curvature at $x = 0$ is

$$\kappa = \frac{1}{(1+0)^{3/2}} = 1$$

The radius if curvature is

$$R = \frac{1}{\kappa} = 1$$

The result is illustrated in Figure 42.

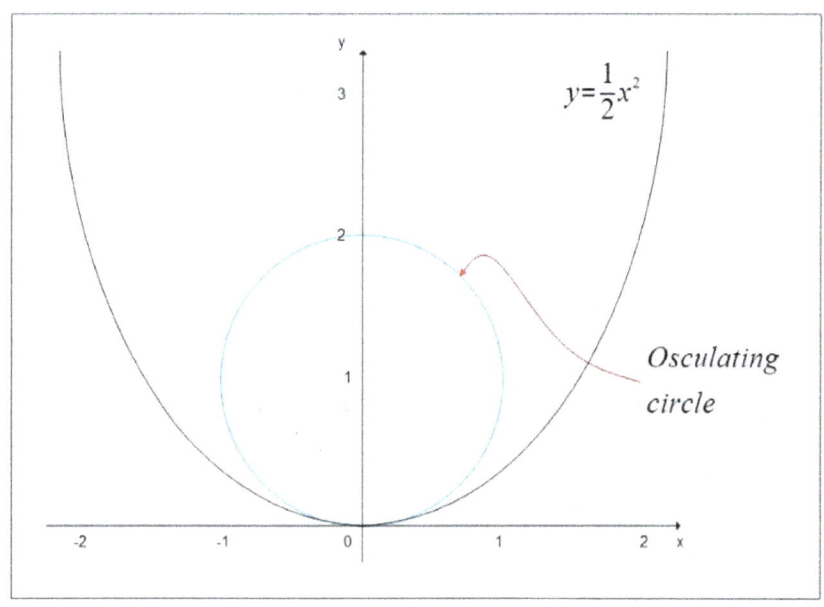

Figure 42 – Curvature of a Parabola

15.4 Integral for a Mean Ergodic Process

Integrate the following:

Kalman Filter

$$I = \int_0^t dx \int_0^t dy \, F(x-y)$$

Make the following transformation:

$$u = x - y, \quad v = y$$

The coordinates of the square in the x-y plane transform to the u-v plane as follows:

$$0,0 \to 0,0$$
$$0,t \to -t,t$$
$$t,0 \to t,0$$
$$t,t \to 0,t$$

The transformation is illustrated in Figure 43.

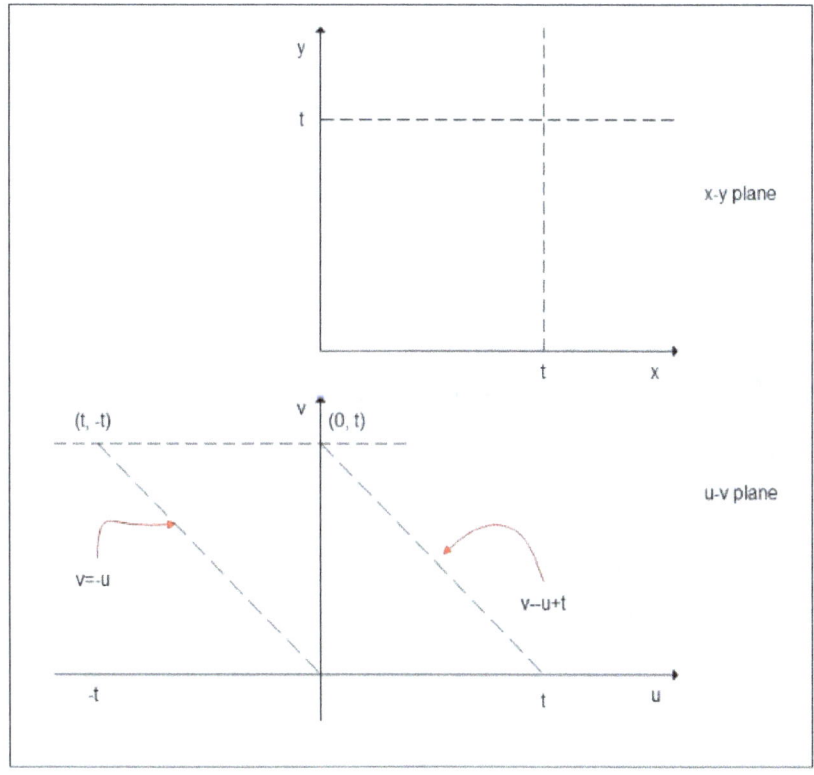

Figure 43 – Change of Variables

Kalman Filter

In the u-v plane, the lines bounding the area to be integrated are

$$v = t$$
$$v = 0$$
$$v = -u$$
$$v = -u + t$$

The Jacobian of the transformation is

$$J = \begin{vmatrix} \frac{\partial u}{\partial x} & \frac{\partial u}{\partial y} \\ \frac{\partial v}{\partial x} & \frac{\partial v}{\partial y} \end{vmatrix} = \begin{vmatrix} 1 & -1 \\ 0 & 1 \end{vmatrix} = 1$$

$$J^{-1} = 1$$

The transformed integral is

$$I = \int_{u=-t}^{u=0} du \int_{v=-u}^{v=t} dv F(u) + \int_{u=0}^{u=t} du \int_{v=0}^{v=t-u} dv\, F(u)$$
$$= \int_{-t}^{0} du\, F(u)(u+t) + \int_{0}^{t} du F(u)(t-u)$$

In the first integral, transform $\alpha = -u$ to obtain

$$= \int_{0}^{t} du F(u)(t-u) + \int_{0}^{t} du F(u)(t-u)$$
$$= 2 \int_{0}^{t} du F(u)(t-u)$$

15.5 Integrals of Odd and Even Functions

When integrating between symmetric limits, such as

$$\int_{-a}^{+a} f(x)\, dx$$

then we have some simplifying results if the function f(x) is of a particular form. The forms of interest are

    symmetric function: $f(x) = f(-x)$
    antisymmetric function: $f(-x) = -f(x)$

The following product rules define multiplication of such functions, where even is denoted by 'e' and odd is denoted by 'o':

Kalman Filter

$$e * e = e$$
$$o * o = e$$
$$o * e = e * o = o$$

Similarly, addition rules are given by:

$$e + e = e$$
$$o + o = o$$
$$o + e = \text{neither o nor e}$$

Any function can be represented as the sum of an odd and an even function. If g(x) is neither odd nor even then:

$$g(x) = \frac{1}{2}[g(x) + g(-x)] + \frac{1}{2}[g(x) - g(-x)]$$

But

$$[g(x) + g(-x)] = [g(-x) + g(x)] \quad \text{even}$$
$$[g(x) - g(-x)] = -[g(-x) - g(x)] \quad \text{odd}$$

and therefore

$$g(x) = e(x) + o(x)$$

In Figure 44 some examples are illustrated. It can be seen that

$$\int_{-a}^{+a} e(x)dx = 2\int_{0}^{a} e(x)dx$$
$$\int_{-a}^{+a} o(x)dx = 0$$

Kalman Filter

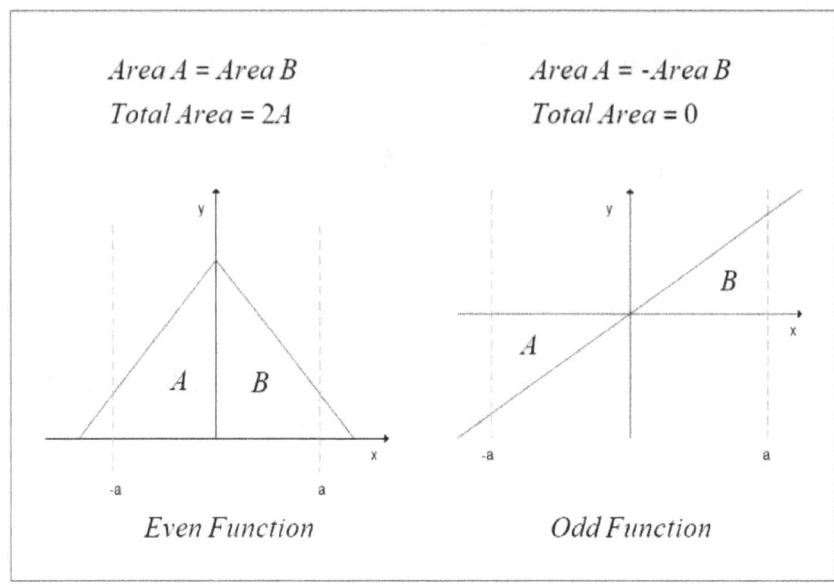

Figure 44 – Odd and Even Functions

15.6 Gaussian Integrals

Gaussian integrals are commonly encountered in DSP and applications of probability. To this end it is useful to become familiar with standard derivations of such quantities. The basic Gaussian integral is given by

$$I = \int_{-\infty}^{+\infty} e^{-ax^2} dx$$

Consider the following product

$$I^2 = \left(\int_{-\infty}^{+\infty} e^{-ax^2} dx\right)\left(\int_{-\infty}^{+\infty} e^{-ay^2} dy\right)$$

$$I^2 = \int_{-\infty}^{+\infty} dx \int_{-\infty}^{+\infty} dy \exp - a(x^2 + y^2)$$

Transform to polar coordinates (and suitable limits):

$$x = r\cos\theta$$

$$y = r\sin\theta$$

$$dxdy = rd\theta dr$$

Kalman Filter

$$I^2 = \int_0^{2\pi} d\theta \int_0^{+\infty} r \exp{-ar^2} dr$$

$$I^2 = 2\pi \left[-\frac{1}{2a} \exp{-ar^2}\right]_0^{\infty} = 2\pi \frac{1}{2a} = \frac{\pi}{a}$$

$$I = \sqrt{\frac{\pi}{a}}$$

It is possible to generate more Gaussian integrals by differentiation of the above result.

Even Functions

Differentiate I wrt a to obtain

$$-\frac{dI}{da} = \int_{-\infty}^{+\infty} x^2 e^{-ax^2} dx = \frac{1}{2a}\sqrt{\frac{\pi}{a}}$$

This method can be continued to derive a series of integrals of an even function of the form

$$x^{2n} e^{-ax^2} \quad \text{For integer n.}$$

Odd Functions

For the integral of the odd function below,

$$I = \int_0^{+\infty} x \exp{-ax^2} dx$$

we have a function and derivative:

$$\frac{d}{dx}\left(e^{-ax^2}\right) = e^{-ax^2}(-2a)x$$

and then

$$I = \frac{-1}{2a}\left[e^{-ax^2}\right]_0^{+\infty} = \frac{-1}{2a}[0 - 1] = \frac{1}{2a}$$

As above, the series can be continued for odd functions by differentiating I wrt to a:

$$\frac{dI}{da} = -\int_0^{+\infty} x^3 \exp{-ax^2} dx = \frac{-1}{2a^2}$$

$$\int_0^{+\infty} x^3 \exp{-ax^2} dx = \frac{1}{2a^2}$$

and so on.

Kalman Filter

16 References

[1] G. Arfken, Mathematical Methods for Physicists, Academic Press.

[2] N. Gershenfeld, The Nature of Mathematical Modeling, CUP, 1999.

[3] M. Schwartz and L. Shaw, Signal Processing: Discrete Signal Analysis, Detection, and Estimation, 1975.

[4] K. W. Behnke and R. A. Singer, "Real Time Tracking Filter Evaluation and Selection for Tactical Operations," *IEEE Trans Aerosp Electron Syst,* Vols. AES-7, no. 1, pp. 100-110, 1971.

17 Index

1/f noise, **42**
Accuracy, **71**
Alpha-Beta Filter, **69**
ARM Cortex, **110**
ARMA, **56**
Autocorrelation, **35**
Autocovariance, **33**
Autoregressive, **56**
AWGN, **85**
Bayes Theorem, **46**
Bias, **71**
Body, **40**
Box-Muller, **112**
Central Limit Theorem, **38**
Chapman-Kolmogorov equations, **65**
Cholesky Decomposition, **106**
Complimentary solution, **22**
Conditional Probability, **29**
Controllable, **83**
Correlation, **34**
Covariance, **33**
Cramer Rao Lower Bound, **77**
Cumulative Distribution Function, **32**
Curvature, **170**
Data Fusion, **166**
Digital Filter, **55**
Dynamics, **12**
EKF, **164**
Emission, **64**
Ergodicity, **45**
Errors, **105**
Estimate, **71**
Estimation, **71**
Exponential distribution, **31**
FIR, **55**
Fisher Information, **75**
Forcing function, **17**
Gaussian, **13**
Gaussian distribution, **38**
Gaussian Integrals, **180**
Hidden Markov Model, **59**
IID, **85**
IIR, **55**

Impulse response, **21**
Innovations, **105**
Inverse Transform Method, **113**
Inversion Relation, **168**
Jacobian, **167**
Johnson noise, **42**
Kalman, **9**, **11**, **14**, **55**, **57**, 65, **69**, **70**, **71**, **74**, **84**, **85**, **87**, 88, **89**, **92**, **94**, **95**, **96**, **98**, **99**, **100**, **101**, **103**, **104**, **105**, **106**, **110**, **118**, **119**, **123**, **127**, **134**, **138**, **139**, **145**, **149**, **150**, **158**, **164**, **166**, **168**
Kalman Filter, **1**, **9**, **11**, **12**, **16**, **21**, **65**, **84**, **88**, **92**, **95**, **96**, **99**, **119**, **127**, **139**, **150**, **164**, **166**
Kalman-Bucy filter, **85**
Laplace transform, **13**
Law of Total Probability, **48**
Least squares, **53**
Leibnitz rule, **19**
Likelihood, **47**, **49**
Linear, **11**, **13**, **17**, **18**, **19**, **20**, **21**, **23**, **41**, **53**, **70**, **74**, **80**, **81**, **82**, **84**, **96**, **100**, **110**, **128**, **138**, **139**, **149**, **150**, **164**, **166**
Linear algebra, **138**, **149**
LU decomposition, **166**
Lyapunov, **24**
Marginal Probability, **29**
Marginalisation, **47**
Markov Chain, **59**
Markov Process, **58**, **59**
Matrix Differentiation, **168**
Matrix Inversion, **104**, **162**
Maximum Likelihood Estimator, **50**
Mean, **29**
Mean Ergodic Process, **176**
Median, **29**
Mode, **29**
Moving Average, **56**
Multivariate Gaussian Distribution, **41**
Noise, **12**, **122**, **132**, **143**, **156**, **161**, **166**
Observable, **81**
Odd and Even Functions, **178**
Optimum, **11**
Particular integral, **22**
Pole, **22**
Positive Definite, **104**
Positive Semi-definite, **104**
Posterior, **47**
Potter, **164**
Precision, **72**
Principle of homogeneity, **18**

Kalman Filter

Prior, **47**
PRNG, **111**
Probability density function, **31**
Probability distribution, **30**
probability mass function, **30**
Random Noise, **84**
Random Process, **42**
Random variable, **29**
Random Vector, **41**
Rank, **80**
Recursive, **11**, **56**, **57**, **73**
Repeatability, **72**
Reproducibility, **72**
Residuals, **105**
Response function, **17**
Roundoff Error, **103**
Shot noise, **42**
Stability, **21**
Standard deviation, **30**
Standardised, **34**
State, **12**, **14**, 61, 80
Stationary, **18**
Statistical Ensemble, **44**
Stochastic matrix, **61**
Stochastic Processes, **58**
Strict Sense Stationary, **43**
Superposition principle, **17**
Tail, **40**
Transition Matrix, **60**
UD factorisation, **166**
Uniform probability distribution, **35**
Variance, **30**
White Noise, **43**
Wide Sense Stationary, **43**